Science Outside the Laboratory

Science Outside the Laboratory

Measurement in Field Science and Economics

MARCEL BOUMANS

Associate Professor History and Methodology of Economics
at Faculty of Economics and Business,
University of Amsterdam and at Faculty of Philosophy,
Erasmus University Rotterdam

OXFORD
UNIVERSITY PRESS

OXFORD
UNIVERSITY PRESS

Oxford University Press is a department of the University of Oxford.
It furthers the University's objective of excellence in research, scholarship,
and education by publishing worldwide.

Oxford New York
Auckland Cape Town Dar es Salaam Hong Kong Karachi
Kuala Lumpur Madrid Melbourne Mexico City Nairobi
New Delhi Shanghai Taipei Toronto

With offices in
Argentina Austria Brazil Chile Czech Republic France Greece
Guatemala Hungary Italy Japan Poland Portugal Singapore
South Korea Switzerland Thailand Turkey Ukraine Vietnam

Oxford is a registered trade mark of Oxford University Press
in the UK and certain other countries.

Published in the United States of America by
Oxford University Press
198 Madison Avenue, New York, NY 10016

Library of Congress Cataloging-in-Publication Data
Boumans, Marcel.
Science Outside the Laboratory : measurement in field science and economics / Marcel Boumans.
p. cm.
Includes bibliographical references and index.
ISBN 978–0–19–938828–8 (alk. paper)
1. Economics—Statistical methods. 2. Social sciences—Statistical methods.
3. Social sciences—Fieldwork. 4. Measurement. I. Title.
HB137.B66 2015
530.801—dc23
2014046362

1 3 5 7 9 8 6 4 2

Printed in the United States of America on acid-free paper

To Freeke

CONTENTS

PREFACE

The first signs were quite alarming: At one of our regular visits at the child health center, one month after the birth of our youngest child, the officiating physician detected cardiac arrhythmia, and so we immediately went to the hospital for further investigation. After all kinds of tests that confirmed the arrhythmia detected by the child health center, the medical staff at the hospital looked even more worried. To investigate how this disorder would evolve and so determine the most appropriate treatment, we received a small box to be connected with stickers to our son's body so an electrocardiogram over a longer period could be taken. At the appointment a week later to discuss the analysis of the ECG, the physician looked surprisingly less worried, even relaxed. The ECG had not shown the arrhythmia again. Carefully examining our son, he commented that he looked very vivid and healthy, and so concluded that no treatment at all was needed. Because the cause of the arrhythmia was—and remained—unknown, the physician could also have prescribed medication as a precaution. This would probably have meant a lifelong dependence on medication. We are happy he did not; our son is now taller than me. The experience reveals that it was the combination of the test results and the clinical view of the physician that induced the right diagnosis.

This book is about what is needed to attain a reliable measurement, particularly when the data out of which the measurement is composed come from different sources, and not all of the different sources of the data derive from clean and controlled environments, which I call "laboratories." One often says that one cannot compare apples to oranges, but this is, however, what one has to do constantly outside these clean and controlled environments, that is, in the "field." To give a simple example, any index number makes such comparisons: take the Human Development Index, where economic growth, life expectancy, and education are "added" together to measure human development.

It is not only that we often need human judgment as one of these required sources of data; the composition of the data has to be designed, and is based on human judgment, too. Mechanical objectivity presumes human judgment to design the mechanism. It is, however, just this human judgment, needed as a source of data and design, that is under attack: evidence is piling up showing how "irrational" and "biased" human judgments are, including those of "experts." In the political domain, one increasingly gets the idea that people need to be "nudged." But who are the "nudgers"? Experts with a special permit? Who gives the permit?

This book is written to explore a methodology of field science to deal with these different sources of evidence, a methodology that aims not at abandoning human judgment but at including it in a "rational" manner. The exploration of this methodology is an exercise in the philosophy of science-in-practice. These practices include both current and past research, where the past covers mainly the twentieth century. In studying various exchanges and practices, I found that those of the 1950s were of most interest and relevance for the subject of this book. It may be that the reason for this is that in that period shortly after the Second World War, with the positive experiences of operations research groups fresh in mind, interaction between various social and natural disciplines was considered to be most promising to achieve progress in the field sciences.

It is a cliché among authors to say that their book could not have been written without the discussions and interactions in the past few years with wonderful minds, but this does not mean it is less true. First of all I would like to express my gratitude to the people of the Center for the History of Political Economy (CHOPE) of Duke University: Bruce Caldwell, Craufurd Goodwin, Kevin Hoover, Neil De Marchi, and Roy Weintraub. By inviting me to join them for the Fall semester in 2012, they provided the intellectual shelter I needed to write the first version of this book. I also would like to thank the visiting fellows of CHOPE during that semester, in particular Verena Halsmayer, with whom I could discuss my first ideas about the overall theme of the book. The manuscript was sent to John Davis, Neil De Marchi, Kevin Hoover, and Mary Morgan for their comments. I am very grateful for their constructive responses, which I used to improve the manuscript. I also would like to thank Giora Hon for his trust and constructive advice, the anonymous Oxford University Press referees for their encouraging and positive reports, and Scott Parris, the Oxford University Press academic book editor, for his enduring trust and support.

This book is based on my work on measurement of the past few years. During this period I benefited enormously from the many intellectual stimulating conversations I had with members of the Amsterdam History and Methodology Research Group, consisting of the late Mark Blaug, Dirk Damsma, Federico D'Onofrio, Harro Maas, Tiago Mata, Mary Morgan, Geert Reuten, Eric Schliesser, and Andrej Svorencik. In particular I would like to thank Freeke Mulder for the many conversations on measurement we had over a much longer period, adequately captured by her motto "Meten is weten," that is, "Measuring is knowing." It is to her I have dedicated this book.

Most of the chapters are derived from articles I have published elsewhere. I thank Duke University Press for permission to use excerpts of my "Observations in a Hostile Environment: Morgenstern on the Accuracy of Economic Observations," *History of Political Economy*, Annual Supplement to vol. 44, 110–131, copyright 2012, Duke University Press, in chapter 1; "The Problem of Passive Observation," *History of Political Economy* 42 (1), 75–110, copyright 2010, Duke University Press, and "Haavelmo's Epistemology for an Inexact Science," *History of Political Economy* 46 (2), 211–229, copyright 2014, Duke University Press, in chapter 4. I thank Elsevier for permission to use excerpts of my "The Role of Models in Measurement outside the Laboratory," *Measurement* 46 (2013) 8, 2908–2912, and "Model-based Type B uncertainty evaluations of measurement towards more objective evaluation strategies," *Measurement* 46 (2013) 9, 3775–3777, in Chapter 2. I thank Sage for permission to use excerpts of my "The Two-Model Problem in Rational Decision Making," *Rationality and Society* 23 (2011) 3, 371–400, in chapter 5. I thank Taylor and Francis for permission to use excerpts of my "The Reliability of an Instrument," *Social Epistemology* 18 (2004), 215–246, in chapter 3, and "Battle in the Planning Office: Field Experts versus Normative Statisticians," *Social Epistemology* 22 (2008) 4, 389–404, in chapter 5. I thank Aksant for permission to use excerpts of my "Measurement and Error Problems (1800–1900): Buys Ballot and Landré's Critique on the Method of Least Squares," in *The Statistical Mind in Modern Society: The Netherlands, 1850–1940*, vol. 2: *Statistics and Scientific Work*, ed. I. H. Stamhuis, P. M. M. Klep and J. G. S. J. van Maarseveen (2008), pp. 179–197, in chapter 3. I thank Springer for permission to use excerpts of my "Model-Based Consensus," in *Experts and Consensus in Social Science: Critical Perspectives from Economics, Sociology, Politics, and Philosophy*, ed. C. Martini and M. Boumans (2014), pp. 49–69, in chapter 6.

Science Outside the Laboratory

1

Introduction

> Overlapping with other social or behavioral sciences—psychology, sociology, history—economics uses the deductive methods of logic and geometry, and inductive methods of statistical and empirical inference. Because it cannot employ the controlled experiments of the physicist, it raises basic problems of methodology: subjective elements of introspection and value judgment; semantic issues of ambiguous and emotional meanings; probability laws of large numbers, both of normal-error-distribution and abnormally skewed type.
>
> —Paul Samuelson, *Economics* (1967, p. 12)

1.1. Science behind the Back of Newton

> Truth is ever to be found in simplicity, and not in the multiplicity and confusion of things. As the world, which to the naked eye exhibits the greatest variety of objects, appears very simple in its internall constitution when surveyed by a philosophic understanding, and so much the simpler by how much the better it is understood, so it is in these visions. It is the perfection of all God's works that they are all done with the greatest simplicity. He is the God of order and not of confusion. And therefore as they that would understand the frame of the world must indeavour to reduce their knowledg to all possible simplicity, so it must be in seeking to understand these visions.
>
> —Isaac Newton[1]

> As far as the laws of mathematics refer to reality, they are not certain; and as far as they are certain, they do not refer to reality.
>
> —Albert Einstein[2]

There is a famous picture of Isaac Newton, painted in 1795 by William Blake, depicting Newton as a "divine geometer." Sitting on a rock, bent over, his sole interest is in his scroll and compass with which he is constructing geometrical forms. Behind his back there is the overwhelming complexity of the natural world. Nature behind the back of Newton is painted in such a way that it seems that Blake wanted to convey that its complexity will never be completely understood by any geometrical or mechanical reasoning. But more alarming is that Newton does not seem to be at all interested in exploring this world. Moreover, this complexity is of such a nature that it cannot be reduced to a specimen that can be studied in a laboratory.

I sympathize with the distressed young girl in a 1768 painting by Joseph Wright of Derby, depicting an experiment on a bird under a bell jar with an air pump. She knows that whatever the demonstration will disclose, it has nothing to do with the pet bird she knew. This book is about science that attempts to acquire knowledge about this complex nature behind the back of Newton and outside the vacuum jar, a science that is called field science. The attempts that will be discussed here to acquire knowledge about these field phenomena particularly relate to measurements.

Although various field sciences will be explored, and the results from these studies are supposed to apply to field science in general, most of the case studies are drawn from the economic sciences. There is no rationale for this, except that economics has such a rich history that it has kept providing me with many rich cases over the many years that I have been studying mathematization, measurement, and modeling in the social sciences.

Measurement is the attempt to acquire quantitative knowledge about a phenomenon. This quantitative knowledge is expressed in numbers. To ensure that these numbers provide reliable information about a phenomenon, specific rules have to be followed. Such rules include a "mapping" of the phenomenon, for example, a model or a formula, plus procedures to ensure that the mapping is reliable in some sense. In other words, and this is the general definition that will be used throughout, measurement is the assignment of numbers to a property of a phenomenon—according to a rule. This book discusses solutions to the following deceivingly simple-looking problem: What are the rules that make these numbers provide reliable information about the phenomenon under study? The solutions discussed are ones that have been developed in the field sciences, but they have also implications for measurement in the laboratory.

But is the measurement of a field phenomenon so different from the measurement of a laboratory phenomenon? Is field science different from laboratory science? To answer these questions, I have to explain what I mean by field science, and what kind of specific problems it engenders for science in general. By field science I mean the enormous and varied range of research practices outside the laboratory. A field science studies phenomena that cannot be studied in a laboratory for practical, technical, or ethical reasons, which means that these phenomena cannot be isolated from their environment and cannot be investigated by manipulation or intervention. But it is not science in the "wild"; before a field phenomenon can be studied, the "environment" has to be "cultivated." By cultivation I mean the installment of scientific institutions, networks, or organizations that enable field study. These institutions are different from laboratories. For example, to acquire meteorological observations a network of observation posts has to be set up.

In an introduction to a special issue of *Osiris* on field science, Henrika Kuklick and Robert E. Kohler (1996) list a number of the problems related to science in the field: As scientific rigor is defined by the standards of the laboratory, the field is considered to be "a site of compromised work: field sciences have dealt with problems that resist tidy solutions, and they have not excluded amateur participants" (p. 1). To discuss science in the field, we will have to take account of a methodological tension between laboratory and field standards of evidence and reasoning.[3]

> We must see how practitioners deal with the difficulties of bringing some order to phenomena that, far more than those of the laboratory, are multivariate, historically produced, often fleeting, and dauntingly complex and uncontrollable. It may seem astonishing, that any robust knowledge comes of fieldwork. Yet it does, abundantly and regularly. (Kuklick and Kohler 1996, p. 3)

While laboratories are exclusively scientific domains, the field is public space. Its borders cannot be rigorously guarded and its population is much more heterogeneous, inhabited not only by scientists but also by other people going about other sorts of business.

Field observations are also more personal than laboratory observations. Take, for example, cases where the field scientist witnesses a singular event. To make such an observation objective, that is, credible and plausible, one cannot rely on the trust-inducing procedures of the laboratory, such as reproducibility, and so the credibility of the observer him- or herself has to be taken into account. But accounting for the credibility of the observer will not make the observation less personal or less subjective. To gain objectivity, institutions have to be created in order to make these personal observations accountable.

The meaning of "personal" used throughout this book comes closest to Leonard J. Savage's (1954) meaning of "personalistic," which he describes as referring to "the confidence that a particular individual has in the truth of a particular proposition," postulating "that the individual concerned is in some ways 'reasonable'" (p. 3). Savage uses this term in order to distinguish it from "objectivistic" and "necessary." "Objectivistic" refers to repetitive events. Here, in this book, this kind of event will be referred to as a laboratory event. Savage uses the term "necessary" to refer to the application of the rules of logic.

There is another relevant difference between a laboratory science and a field science: A field science is much more inexact than a laboratory science. What the difference is between an exact and an inexact science will be explicated in more detail in chapter 4, but for now it suffices to clarify it by comparing the semantics of the equals sign "=" in an exact science with its semantics in an inexact science.

In an exact science, the equals sign means exact equivalence of what is expressed on both sides of the equals sign of the "equation." In an inexact science, the equals sign means that what is expressed on the right side implies what is expressed on the left side of the "relation." This general meaning of the equals sign as an implication includes various specific meanings, such as logical implications and causal connections. Whatever its specific meaning, the equals sign in an inexact science does not necessarily mean symmetry, unlike the equals sign in an exact science.

When the equals sign denotes a causal relation in an inexact science, it means that the right side of the relation is not expected to be complete, that is, the right side does not contain the complete set of all possible factors that are causally connected to what is represented on the left side. In an exact equation it does. Newton's law $f = m \cdot a$ (force f is mass m times acceleration a) is an exact expression, but the law of demand, $D = -c_1 p + c_2$, (demand D is negatively (cor)related to price p) is not exact. The inexactness of economic laws will be discussed in more detail in chapter 4, but the inexactness alludes to the incompleteness of economic theories. Field phenomena are usually not simple; presumably a large number of causal factors are involved, not all of which can be measured or are even known. These two themes, inexactness and incompleteness, are key issues when dealing with measurement in field science.

To clarify the above problems of field science, I will use Oskar Morgenstern's (1963a) *On the Accuracy of Economic Observations* as an illustrative case for discussing the reliability of field observations. Morgenstern's book discusses extensively the "accuracy" of field observations, by exploring the various sources of errors, how to deal with, reduce, or—if at all possible—avoid them.

1.2. Oskar Morgenstern on the Accuracy of Economic Observations

The work of Oskar Morgenstern, a twentieth-century economist, is for several reasons an appropriate case to introduce the topic of this book. The 1940s and 1950s were a period in which methodologies for science outside the laboratory were intensively explored and discussed in psychology, economics, and philosophy. Morgenstern was present at several events and affiliated with organizations where the discussions and developments of these methodologies later came to be considered important and even crucial.

Morgenstern is best known for his 1944 book, coauthored with John von Neumann, *Theory of Games and Economic Behavior*. The axiomatic framework they had set up for the measurement of utility was highly influential in the development of the representational theory of measurement (see chapter 2). At the Rand Corporation, he interacted with Olaf Helmer, the creator of the Delphi method, on the inexactness of social science and the need for the involvement of expert judgment (see chapter 7).[4] And he participated in an eight-week seminar, "The Design of Experiments in Decision Processes," that was held in Santa Monica in 1952 to accommodate the game theorists and the experimenters associated with the Rand Corporation. This event has been viewed as the birthplace of experimental gaming (Innocenti and Zappia 2005, p. 80), but it can also be considered the birthplace of a measurement theory for social science (see chapter 2).

Morgenstern had a lifelong interest in economic statistics and in determining the best scientific methodology for dealing with them. In a letter to Eve Burns (dated 2 March 1928) Morgenstern stated, "I deal extensively with the question of quantitative and qualitative methods and with the statistification [*sic*] of economic theory etc. . . . You say that you study science. This is very laudable, but again I want to warn you not to forget, that the method and aims of our social sciences are very much different from the natural sciences" (quoted in Mirowski 1992, p. 130). In 1931 he succeeded Friedrich Hayek as director of the Vienna Institute for Business Cycle Research, a position that he held until his move to America in 1938. He became an elected member of the International Statistical Institute in 1937 and maintained membership until 1973.

Morgenstern's early experiences with statistics were reported in his *Wirtschaftsprognose* (1928). In a review of this book, Arthur Marget summarizes Morgenstern's methodology of economic forecasting as three propositions:[5]

1. Forecast in economics by the methods of economic theory and statistics is "in principle" impossible.
2. Even if it were possible to develop a technique of economic forecasting, such a technique would be incomplete, by virtue of its necessary limitation to methods based on a knowledge of economics alone; it would therefore be incapable of application in actual situations.
3. Moreover, such forecasts can serve no useful purpose. All attempts to develop a formal technique for forecast are therefore to be discouraged. (Marget 1929, p. 313)

Marget then distinguishes three "subpropositions" that support the first proposition:

a. The data with which the economic forecaster must deal are of such a nature as to make it certain that the prerequisites for adequate induction, that is, the application of the technique of probability analysis, must always be lacking.
b. Economic processes, and therefore the data in which their action is registered, are not characterized by a degree of regularity sufficient to make their future course amenable to forecast, such "laws" as are discoverable being by nature "inexact" and loose, and therefore unreliable.
c. Forecasting in economics differs from forecasting in all other sciences in the characteristic that, in economics, the very fact of forecast leads to "anticipations" which are bound to make the original forecast false. (Marget 1929, pp. 313–314)

Marget's review summarizes not only a very early work by Morgenstern but also the two themes Morgenstern continued to study in his later works on economic statistics, namely,

1. The inadequacy of the statistical approach.
2. The looseness and inexactitude of economic laws.

The problem of inaccuracies of economic observations was a topic Morgenstern had to deal with even more prominently when he was, from 1936 to 1946, a member of the Committee of Statistical Experts of the League of Nations.[6] As a member of this committee, he had to deal with the main problem of international statistics, namely, comparability.

> Summarizing, we can state that statistics giving international comparisons of national incomes are among the most uncertain and unreliable statistics with which the public is being confronted. The area is full of complicated and unsolved problems, and in spite of the great efforts to overcome them, the progress is slow. This is a field where politics reigns supreme and where lack of critical appraisal is particularly detrimental. (Morgenstern 1963a, p. 282; see also 1963b, p. 173)

For example, in the case of international foreign trade statistics, there are many reasons for divergence in the data relating to the same trade flow in statistics of partner countries. These reasons can be grouped under three headings (see Federico and Tena 1991, pp. 261–263). The first is "unavoidable" differences arising between nonbordering countries because of the time and cost of transportation. The second is "structural" differences in compilation criteria, which could be eliminated by standardization. The third is actual errors, that is, cases where recorded data differed from the real flow; they can be classified as follows: (1) failure to record because of smuggling; (2) inaccurate recording following wrong declarations because of negligence or fraud; and (3) errors by statistical offices.

A discussion of the accuracy of economic observations is a discussion of the specific nature of economic data, a subject that "always occupied a central role" (Morgenstern 1963a, p. vii) in Morgenstern's work, but was discussed most extensively in *On the Accuracy of Economic Observations* (1963a).

The book's main message was that in comparison with data used in the natural sciences, economic and social statistics have additional peculiarities: "at least all sources of error that occur in the natural sciences also occur in the social sciences" (p. 7).

Morgenstern argued that the accuracy of economic observations cannot be "formulated according to a strict statistical theory for the simple reason that no such exhaustive theory is available for many social phenomena" (p. 7).

Therefore the treatment of errors in economic data has to be a "common-sense" approach. The reason for not being able to apply statistical theory is that the nature of economic data prevents "normal distribution of the observations, creating circumstances which cannot be readily treated according to classical notions of probable error" (p. 13):

> The notion that errors do cancel out is widespread and when not explicitly stated, it appears as the almost inevitable argument of investigators when they are pressed to say why their statistics should be acceptable. Yet any statement that errors "cancel," neutralize each other's influence, has to be proved. Such proofs are difficult and whether a "proof" is acceptable or not is not easy to decide. The world would, indeed, be even more of a miracle than it is if the influence of one set of errors offsets that of another set of errors so conveniently that we need not to bother much with the whole matter. (Morgenstern 1963a, p. 53)

Morgenstern (1963a) listed 10 sources for errors in economic statistics:

1. Lack of designed experiments
2. Hiding of information, lies
3. The training of observers
4. Errors from questionnaires
5. Mass observations
6. Lack of definition or classification
7. Errors of instruments
8. The factor of time
9. Observations of unique phenomena
10. Interdependence and stability of errors

These sources of errors were discussed by Morgenstern in comparison with natural science observations. Economic statistics are not the result of designed experiments, and besides they are often dependent on legal rather than economic definitions of processes. Moreover, even when planned, economic statistics are generally not gathered by "scientific observers." "A scientific observer is the astronomer at his telescope, the physicist recording the scatter of mesons, the biologist determining the hereditary behavior of some cells, etc.; all are themselves scientists; they do not operate through agents many times removed" (Morgenstern 1963a, p. 27).

Because of the enormous amount of data needed in economics, "scientific observations" would be physically impossible: "We cannot place technically trained economists or statisticians at the gates of factories in order to determine what has been produced and how much is being shipped to whom at what prices. We will have to rely on business records, kept by men and, increasingly, by machines, none of them are part of the ideally needed scientific set-up as such" (p. 27).

What Morgenstern saw as a main source of errors in economics, namely, that most observations were nonscientific, was emphasized once more in his 1959 book *International Financial Transactions and Business Cycles*:

> Economic statistics are—in the overwhelming majority of cases—*not scientific observations*. This is a point of primary significance. They are at best historical accounts; mostly they are byproducts of business operations or of administrative acts. They are, as a rule, badly collected by scientifically untrained minor officials at the customhouses, warehouses, on street markets, etc. In other words they are not the results of carefully set experiments, or of strictly controlled measurements as are astronomical observations. (Morgenstern 1959, p. 9)

Many economic observations concern events that are unique and not reproducible. One is usually confronted with historical processes. In the case that a unique event is observed more or less simultaneously by several independent but differently placed observers, one has to consider the following problem:[7]

> It is then necessary to decide which one is to be trusted (with his own observational errors still remaining), or whether averages are to be taken, what kind of averages, etc. . . . The same problem occurs occasionally also in physics and astronomy, e.g., in the field of extraordinary sound propagation, the measurement of explosions, the accounts of eruption of volcanoes, spring tides, and the observation of novae. Statistical theory adequate for full consideration of all issues raised under these conditions apparently does not exist and will be difficult to develop. (Morgenstern 1963a, p. 48)

The most "profound" difference between natural science data and social science data, however, is that the latter are "frequently based on evasive answers and *deliberate lies* of various types. These lies arise, principally, from misunderstandings, from fear of tax authorities, from uncertainty about or dislike of government interference and plans, or from the desire to mislead competitors" (p. 17).

Nature may hold back information, may be difficult to understand, but "she does not lie deliberately" (p. 17). To clarify what he meant by this, Morgenstern refers to Albert Einstein's famous pronouncement: "Raffiniert is der Herr Gott, aber boshaft ist er nicht," inscribed on the mantle of a fireplace in Fine Hall in Princeton University, and which Morgenstern translated as follows: "The Lord God is sophisticated, but not malicious."[8] Morgenstern emphasized that there is "a significant variation in the structure of the physical and social sciences, provided it is true that Nature is merely indifferent and not hostile to man's efforts to finding out truth—it certainly not being friendly" (p. 18).

This was not a loose remark: in at least two research papers, one written in 1951 and the other in 1966, Morgenstern referred explicitly to Nature's benevolence. In the 1951 paper "Prolegomena to a Theory of Organization" (Morgenstern 1951), he introduces Nature's benevolence when discussing the signaling system of an organization. For a signaling system of an organization there is no distinction between event and signal: an event becomes known only through a signal, so for an organization they are inseparable. Morgenstern classified the events relevant to an organization into "events of the organization" and "other events" that are either "physical, i.e., produced by Nature" or "other organizations' choices" (p. 55). "Events of Nature" are determinable if their probability distributions are known. But even if their probabilities cannot be estimated, Morgenstern emphasizes that "Nature is never malevolent, i.e., bent on impeding the organization in the pursuit of its aim" (p. 55).

Because the social world is neither benevolent nor simple, the assumptions on which the theory of error is based (see chapter 3 for what these assumptions are) are not appropriate to deal with social statistics. Put another way, statistical theory, mainly developed to deal with natural phenomena, is incomplete with respect to social phenomena.

In a 1966 paper, Morgenstern elaborated in more detail on "Nature's attitude" to discuss its epistemological consequences: While man's behavior is "mostly hostility," "Nature, however, may be benevolent, or at least indifferent, to us. Nature is generally not considered malevolent to man" (p. 8).

> If nature is indifferent to man or even benevolent, we may proceed with our methods as we have done; but if there is a suspicion of hostility, our approach would become most difficult indeed. Instead of using pure, direct strategies in questioning nature we would have to develop different ideas, perhaps similar to those needed in social science, where often an indirect approach is needed in order to elicit the truth from the subjects studied and to avoid the contamination of the observer due to his immersion in the society he studies. (Morgenstern 1966, p. 18)

Deliberately untrue statistics offer a most serious problem with broad ramifications in the realm of statistical theory. But while the nature and consequences of such statistics, according to Morgenstern, had not been explored sufficiently, he nevertheless noted that "a theory of 'sampling in a hostile environment' is now under development" (Morgenstern 1963a, p. 21). The development of such a theory was seen by Morgenstern in the context of game theory, the setup of a nonstrictly determined two-person game where both sides have to resort to mixed or "statistical strategies": "It is an ironic circumstance that in order to get good statistics, 'statistical strategies' may have to be used!" (p. 22).

Morgenstern emphasized that to get good statistics, it is important "to understand that there is a fundamental difference (in the field of economics) between mere *data* and *observations*" (p. 88). Observations are planned, designed, and guided by theory, such as, for example (and not necessarily), those obtained in a controlled experiment. Data are merely obtained, gathered, and collected statistics, even though this involves administrative planning. To explain how he saw the difference between statistics, data, and observations and their relation to theory, Morgenstern provided the Venn diagram shown in figure 1.1.

A is the body of data in the above sense, **B** represents "other data, such as historical events or non-measurable data," and **C** is the theory "partly based on **A** and **B**." Morgenstern defines "scientific information" as the intersection of **A** and **C** ("quantitative information"), the intersection **B** and **C** ("description"), and the intersection of **A**, **B**, and **C** ("observation"). Scientific information is thus always related to theory. Most economic data, according to Morgenstern, are of the class **A** minus **C**: "These data as such tell no story" (p. 89).

While his 1950 book on observations did not further explore this distinction between data and observation, two years later Morgenstern provided a fuller account of this distinction in a paper on experiment and computation in economics. The paper was presented at the above-mentioned Rand seminar, "The Design of Experiments in Decision Processes." Thirty-six papers were presented. A smaller number of the papers (nineteen) were published in *Decision Processes* (Thrall, Coombs, and Davis 1954), of which only six were reporting results of experiments (Roth 1993, p. 194). Both authors of *The Theory of Games* (1944) presented a paper, which did not appear in *Decision Processes,* and remarkably were about computers: John von Neumann's "Remarks on Chess-Playing Automata" and Morgenstern's "Experiment and Computation in Economics." The latter paper was published in 1954 in a collection of papers edited by Morgenstern on economic activity analysis and linear programming (Roth 1993, p. 194), now with a slightly extended title: "Experiment and Large Scale Computation in Economics." Morgenstern discussed the relation of experiment to computation, that is, computational experiments, which did not involve, or at least not explicitly, decision processes or game theory.

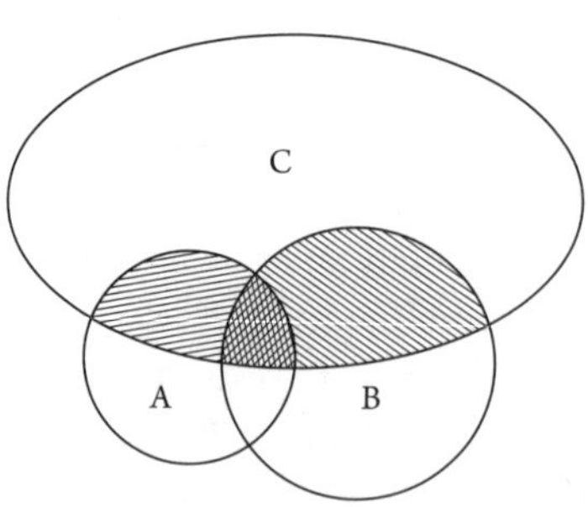

Figure 1.1 Morgenstern's diagram showing the difference between data, information, description, and observation. Source: Morgenstern 1963a, p. 89. "On the Accuracy of Economics Observations," Oskar Morgenstern Papers, David M. Rubinstein Rare Book & Manuscript Library, Duke University.

Morgenstern's (1954) starting point is that experiment and computation are closely related. To clarify this relation, he distinguishes between two ways of using computations: first, as substitutes for experiments, and second, to generate new data. Elaborating on this, he subsequently distinguishes two types of experiments:

1. Experiments of the *first kind* are those where new general properties of a system are to be discovered by its manipulation on the basis of a theory of the system;
2. experiments of the *second kind* do not primarily rely on a theory but aim at the discovery of new, individual facts. (Morgenstern 1954, p. 499)

Using these definitions, Morgenstern's main "thesis" was the following: "Every computation is equivalent to an experiment of the first kind and vice versa" (p. 499). This equivalence was, according to Morgenstern, already emphasized by Ernst Mach through his notion of thought experiment.

> Its methods involve imagining conditions that differ from the known conditions and then attempting to identify the proper factor to which the imagined variations could be ascribed. This procedure consists in the drawing of implications and like other experiments may lead to the discovery of new facts. (Morgenstern 1954, p. 486)

Thought experiments were "vitally affected" by the new possibilities created by computers.

Subsequently, Morgenstern also equated "planned and controlled observation or measurement" with experimentation: measurement is "at least an experiment of the second kind, but may be one of the first kind. This implies that there is no sharp dividing line between experiment and measurement" (p. 506). Thus, according to Morgenstern, there is no principal distinction between a physical experiment, "one in which physical reality is being subjected to desired conditions" (p. 486), and planned observation.

> The definition of an experiment as a process in which primarily a change in the direction of forces or the like is considered would definitely be too narrow if sensible at all. The largest part of scientific activity which is unmistakably experiment would be left out of account. (Morgenstern 1954, p. 508)

Experiment, observation, and measurement are so intricately interrelated that a distinction between observational sciences, such as astronomy, and experimental sciences, such as physics, is doubtful.

> It is true that the astronomer cannot change the course of stars and in that crude sense cannot make an experiment (i.e., make those observations

> and measurements which such interferences would make possible). But he does approach the multiple phenomena of the stars experimentally, for example by the use of telescopes, photography, spectroscopy, etc. That is, the experiment consists in approaching the basic phenomenon by means of numerous devices thereby improving its description. . . . The devices used in astronomy are in no way different from those applied to conditions in a laboratory. (Morgenstern 1954, pp. 507–508)

Likewise, in economics, planned observation is principally not different from experimentation:

> It is enough to have to count anything at all (let alone the population of a whole nation) or to make a simple enumeration of the units of some good produced, in order to be confronted with the entire array of questions that have to be considered when setting up an experiment in the conventional sense. There must be "strict control" of as many parts of the process of counting as it is possible to achieve. When the counting requires more than a piece of paper and a pencil of one man, the social scientist, precisely as the physical scientist, is immediately involved with the interaction of men and machines of all kinds from the simplest abacus to electronic IBM equipment. All these involve as much "experiment" as anyone may wish to have to deal with, even though the name is not customarily attached to the procedure. (Morgenstern 1954, pp. 506–507)

For planned observations to be scientific, however, Morgenstern emphasized the essential role of theory: "If there were no theory behind the experiment but still the intent to discover general properties of a system, there would be no experiment at all, only a meaningless muddling" (p. 502). In the first place, theory reduces the amount of required measurements: "The better the underlying theory the fewer will the direct experimental trials (measurements) have to be and the more weight can be thrown on the computations (at a given technology of computing)" (p. 509). This specific role of theory was also mentioned in his 1959 book on international financial statistics:

> *The amount of statistical data* needed depends on the state of the theory that can be used in the exploration of a field. The better the theory, i.e., the fewer the principles it uses, the fewer additional statistics are needed to test or advance it. (Morgenstern 1959, p. 11)

In the second place, theory is needed to attach any meaning to data. Without theory, one would be "just looking" or "merely looking." In an early stage of science this may lead to data of a new kind. When the telescope and the microscope were invented, "all that mattered was to take these wonderful new instruments and to look, to look practically anywhere. Some phenomena would turn up, totally unsuspected, be they the moons of Jupiter or some tiny amoebae in a drop of water" (p. 540). But "one would not get very far today either in astronomy or biology if one were merely to look around as one did then: now theory is guiding the direction of the search for facts and phenomena, sometimes for exceedingly esoteric and elusive ones" (p. 540).

This situation where new instruments lead to data of a new kind by merely looking is, according to Morgenstern, the current situation of economics. Comparable to the telescope and the microscope, we have the "high-speed electronic computer." But this new possibility of processing enormous amounts of data is, according to Morgenstern, still "mere looking." Though "the field for 'mere looking' in biology having practically vanished," in the "comparatively undeveloped state of economics the 'mere looking' by means of computers occupies still a very large field for the future" (p. 542).

So to Morgenstern "scientific observation" meant experimentation of the first kind, that is, not merely looking around but theory-guided planned observation. But unfortunately—according to Morgenstern (1963a, p. 88)—in economics, theory is not exact, and is never based solely on data; it is "constructed and invented" and to "a very high extent in addition related to non-constructively obtained material, such as personal experience." This lack of exact theory in economics was the main reason for Morgenstern to be pessimistic about improving the accuracy of social statistics.

Because of the state of economics, that is, its incomplete and inexact theory, Morgenstern's (pessimistic) view was that scientific observation in the sense of theoretical experiment is hard to achieve in the discipline. Because the alternative of "just looking," however, is too vulnerable for deceit, Morgenstern suggested the involvement of scientific observers having appropriate knowledge, even though their knowledge may be inexact. In a field that does not have explicit and exact theories, a scientific observation is an observation made by an scientific expert having intuitive knowledge of the relevant phenomena.

When drafting the outlines for a theory of "sampling in a hostile environment," Morgenstern had a specific role for "scientific observers" in mind. As a member of the Committee of Statistical Experts of the League of Nations, he had to deal with hostile governments, like that of Nazi Germany in the 1930s and later, being affiliated with Rand in the 1940s and 1950s, with those of the Soviet Union and China.

> A special problem is offered by the Soviet Union. The statistics of that country are exceedingly difficult to assess, but it is generally known that they are seldom what they purport to be. There has been a great deal of deliberate doctoring of statistics at many levels, in order, for example, to make production results appear better than they are or to receive assignments of raw materials that would not otherwise be allocated. Even Khrushchev has repeatedly referred to falsified accounts of various

> activities, especially in farming, and there is no reason to assume that statistical practices were better in Stalin's time. (Morgenstern 1963b, p. 173; see also 1963a, p. 280)

A scientist, in this case an economist, with knowledge of the Soviet economy, would be able to see through the "doctored" data a glimpse of the real economic situation in the Soviet Union.

According to Morgenstern, for an observation to be scientific, it should be planned, designed, and guided by theory. He compared scientific observation with experimentation, but in contrast to physics, in economics the theoretical guidelines are inexact and more intuitive. Because economic theory is inexact and incomplete, we need experts to assess the supplied economic statistics.

These experts are necessary for another reason, beyond the inexactness of economic laws. Morgenstern, when discussing the accuracy of economic information, is well aware of the distinction between observations of nature and observations of business, institutions, and governments. Observations of natural phenomena can be inaccurate, but the only one to blame for this is the observer: Nature does not lie. In contrast, a human agency providing information can also be the cause of inaccuracies, sometimes deliberately. A scientific observer, however, because of his or her knowledge of the "laws" of economics, may be able to see whether a picture based on economic data is diverging from a "natural" picture.

1.3. A Social Science Perspective

While Morgenstern nicely enumerates the problems one has to deal with in a field science, such as inexactness of theories, incompleteness of statistical theories, unreliable statistics, and lack of "scientific" observations, his position was atypical. It is atypical because he took the natural science, that is, laboratory science, approach as the ideal standard for dealing with errors in contrast with contemporary social statisticians.

In a comment on Morgenstern's (1963b) article in *Fortune*, in which Morgenstern had summarized the main concerns of the 1963a edition of *On the Accuracy*, Raymond T. Bowman (1964, p. 10) noted that recognizing and emphasizing the need for attention to error will not lead to an "uncritical discrediting" of economic and social statistics.[9] According to Bowman, Morgenstern set up the natural sciences "as the ultimate standard by which the achievements of the social sciences should be judged" (p. 10).

To evidence his view on Morgenstern's position, Bowman referred to a comment that Simon Kuznets made in a review of the 1950 edition of *On the Accuracy of Economic Observations*. Kuznets, who had presented, on behalf of Morgenstern, the abstract of *On the Accuracy* at the twenty-sixth session of the International Statistical Institute, which took place in September 1949

(Morgenstern 1963a, pp. v–vi), had also written a review of the book (Kuznets 1950b). Because of Kuznets's involvement with the US national income accounts, his discussion on inaccuracies of economic observations implied a different direction for dealing with errors than Morgenstern had proposed in his book.

Kuznets argued as follows: Economic statistics are products of changing social institutions and relate to changing historical reality. Errors in such data are, therefore, complex and largely unique historical phenomena, "which is but another way of saying that we are not dealing here with the results of designed, controlled experiments" (p. 577). Lack of attention paid to errors in economic statistics compared to those data in the natural sciences may have occurred because of "a feeling of helplessness and a realization of the difficulties involved in dealing with them effectively" (p. 577). Economic statistics have to be seen as records of institutional conventions rather than experimental conventions, "as defined under imaginatively controlled conditions" (p. 577). Therefore Kuznets suggested dealing with accuracy by the rules of accounting. Moreover, in research or in policymaking, one does not rely on a single series but on a variety of data. Error evaluations are thus based on a consensus of these various data. According to Kuznets, Morgenstern tends in his discussion "to set up the natural sciences as a feasible ideal" (p. 579), so as to understate that economic statistics are records of constantly changing institutional settings. But Kuznets also noted that Morgenstern is quite aware of the principal difference between natural and social data:

> The reader might too easily conclude from Professor Morgenstern's discussion that the trouble lies largely in the lack of attention paid by economists and statisticians to the problem; and that increased attention would go far towards solving the problem, as it has been solved in the natural sciences. To such a conclusion the reviewer, and most likely also the author, would enter a strong objection. (Kuznets 1950b, p. 579)

In his presidential address to the American Statistical Association, Kuznets (1950a), like Morgenstern, made a similar distinction between observations in natural science and in social science and its consequence for the reliability of the statistics:[10] in natural science, "where controlled experiment is possible, the experimenter produces his own data—in accordance with his analytical goals," whereas social statistics are produced by various agencies, and "the production of the data is a *social*, not an individual, act" (p. 2).

> The views of these agencies necessarily differ from those that would be entertained by a scientifically minded analyst, even one who happened to live at the same time and in the same place. Consequently, the supply of data is capricious as judged by any consistent and reasoned standard for scientific analysis. (Kuznets 1950a, p. 3)

Hence the social origin, which is acquired in an experimentally uncontrolled way, of social statistics affects, according to Kuznets, their reliability.

Errors may arise from three different sources: lack of control by the respondent, lack of control by the collecting agency, and lack of control by the "analytically minded final user." Errors from the respondents "may easily arise either because they deliberately falsify or because their knowledge is not full or accurate" (p. 4). All three problems can be minimized in controlled experiments where the observations are made by the experimenter self.

Although both Bowman and Kuznets shared Morgenstern's concerns with respect to the reliability of social statistics, they did not share his pessimism about improving its reliability, because they did not share his natural science ideal. This ideal kept Morgenstern from developing a methodology for achieving more accuracy. Although Morgenstern had paid sufficient attention to the problem of errors in economic statistics, Bowman was disappointed that Morgenstern's work had contributed "little to proposing the framework or the practical procedures which might advance the work now going forward on the reduction of statistical errors" (Bowman 1964, p. 19).

I also do not share Morgenstern's pessimism. This book attempts to outline a methodology for an inexact science that is based on the standards of social science. This implies that we will have to reconsider issues of objectivity and rigor, which are often defined from a natural science, that is, laboratory science, perspective. It also implies that we will have to reconsider the treatment of observations.

A treatment of observations is a treatment of sources of observational errors, which I call the "calculus of observations" (see chapter 3). A mathematical theory of error could arise only when observational errors were divorced from cause and effect, from the individual observer, from actual measurement, from time, and so forth (Klein 1997). Abstraction from these idiosyncratic circumstances led to the mathematical theory of error based on the works of Carl Friedrich Gauss, Adrien-Marie Legendre, and Pierre-Simon Laplace.[11] With key elements as the least squares method and the normal distribution, this theory is based on the assumption that by averaging the observations, the observational errors will cancel each other out. This is a history of creating objectivity: the elimination of personal judgment by mechanical rules of calculation. It was meant to "filter out local knowledge such as individual skill and experience, and local conditions such as this brand of instrument or that degree of humidity" (Daston 1995, p. 9). It will appear that for reliable measurements of field phenomena these filtered-out elements have to be accounted for, or even stronger, are needed.

The history of the theory of error is a history in which the errors were attributed elegant bell-shaped characteristics that enabled mathematization and the creation of objectivity. According to this theory, it is not a problem that observations are made by laypeople, because this theory tells how personal biases can be neutralized mathematically. The underlying assumption is that Nature is benevolent and simple. The social world, however, can be hostile, more secretive, and complex. Attributing simple and nice characteristics would not do. It requires the development of a theory of sampling in an unfriendly environment. Getting information about such a world asks for scientists who are experts in hidden and tacit regularities and are able to distinguish between counterfeit and fact. A history of field observations is a history of subjectivity, scientific intuition, and idiosyncrasy.

1.4. Experimentation

There seems to be a broadly shared consensus about which epistemic genre can be considered as most scientific, namely the experimental method, and more precisely, the "laboratory ideal" of experimentation, that is "designing manipulated, well-controlled, isolated experimental systems" (Schwarz and Krohn 2011). Although this ideal is often presented as the official position, Astrid Schwarz and Wolfgang Krohn (2011) observe that the actual practice of experimentation has shifted towards the field ideal of experimentation and that this shift has been hardly addressed by philosophers of science. Notwithstanding this general lack of attention to field experimentation, in the 1950s, however, it was discussed among philosophers of science, mainly with respect to the possibilities of experimentation in the nonlaboratory sciences.

Since the nineteenth century, scientists and philosophers are aware that many natural and social phenomena cannot be studied in a laboratory (see also chapter 4). For these phenomena the most prominent alternative method since the 1950s became the method of statistical modeling. Instead of experimenting on real phenomena, one could experiment on the phenomena existing in the world of the model (Morgan 2012). Both illustrative as well as representative for this general understanding are the comments by Olaf Helmer in a 1953 internal Rand Corporation memorandum "Experimentation in the nonexperimental sciences"[12]: "in economics and political science it is our unwillingness and, often, inability to interfere with social institutions as well as the seeming impossibility of achieving the prerequisites of all experimentation, the reproducibility of like circumstances" (Helmer 1983, pp. 23–24). Helmer suggested experimentation on an analogue system when direct experimentation on a phenomenon was not feasible.

> In short, the real world is first replaced by an analogue (usually mathematical, sometimes physical), the experiment with real objects is replaced by one with fictitious objects, and in this "experiment" the features supposedly not under the control of the "experimenter" are assumed to be subject to given probability distributions, so that the chance fluctuations that would naturally occur can be simulated in the model by the operations of an artificial chance device. (Helmer 1983, p. 24)

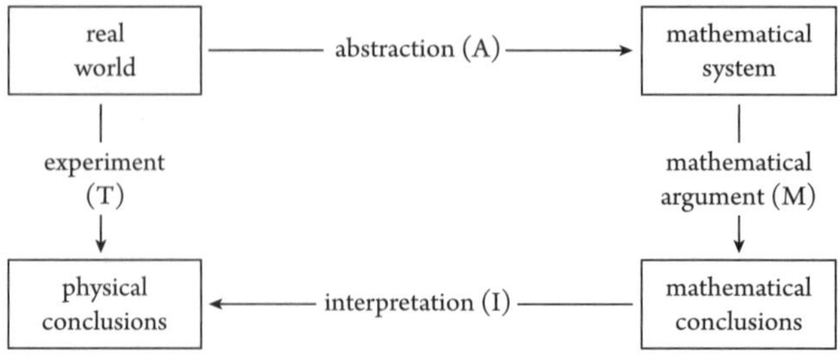

Figure 1.2 The symmetrical roles of experiment and model. Source: Coombs, Raiffa and Thrall 1954a, p. 133.

Helmer's view was typical of the 1950s. Another example of this view is the one presented in a paper, "Some Views on Mathematical Models and Measurement Theory," by Clyde H. Coombs, Howard Raiffa, and Robert M. Thrall (1954a, 1954b).[13] In this paper they argued for "symmetrical roles" of experiment and inferences from a mathematical model, called the "mathematical argument."[14] Their view of the role that mathematical models play in science is illustrated in figure 1.2.

The role of a mathematical model as alternative to experimentation was clarified as follows:

> With some segment of the real world as his starting point, the scientist, by means of a process we shall call abstraction (*A*), maps his object system into one of the mathematical systems or models. By mathematical argument (*M*) certain mathematical conclusions are arrived at as necessary (logical) consequences of the postulates of the system. The mathematical conclusions are then converted into physical conclusions by a process we shall call interpretation (*I*). (Coombs, Raiffa, and Thrall 1954a, pp. 134–135)

The figure expresses clearly the authors' view on science and the "symmetrical roles" mathematical models and experiments have in science[15]:

> The task of a science looked at in this way may be seen to be the task of trying to arrive at the same conclusions about the real world by two different routes: one is by experiment and the other by logical argument; these correspond, respectively, to the left and right sides of [figure 1.2]. There is no natural or necessary order in which these routes should be followed. (Coombs, Raiffa, and Thrall 1954a, pp. 134–135)

Which route to be taken depends on the subject under study. The process of measurement, corresponding to *A* in figure 1.2, was according to Coombs, Raiffa, and Thrall (1954a, p. 136) "an excellent illustration of the role of mathematical models."[16]

The importance and relevance of using homomorphisms in science was emphasized by the rise of "transdisciplinary" approaches like "cybernetics" and "systems theory" in the 1940s and 1950s, and was stimulated by the successful experiences with "operations research" in World War II. In an early paper, even before the term "cybernetics" was used (by Norbert Wiener), Arturo Rosenblueth and Wiener (1945) discussed the role of models in science. They saw the intention and result of scientific inquiry to be obtaining understanding and control. But this could only be obtained by abstraction:

> Abstraction consists in replacing the part of the universe under consideration by a model of similar but simpler structure. Models, formal or intellectual on the one hand, or material on the other, are thus a central necessity of scientific procedure. (Rosenblueth and Wiener 1945, p. 316)

They distinguish between material and formal models:

> A material model is the representation of a complex system by a system which is assumed simpler and which is also assumed to have some properties similar to those selected for study in the original complex system. A formal model is a symbolic assertion in logical terms of an idealized relatively simple situation sharing the structural properties of the original factual system. (Rosenblueth and Wiener 1945, p. 317)

Material models are, according to the authors, useful in two ways. Firstly, "they may assist the scientist in replacing a phenomenon in an unfamiliar field by one in a field in which he is more at home" (p. 317). Second, a material model "may enable the carrying out of experiments under more favorable conditions than would be available in the original system" (p. 317).

This idea that models could be used when experiments were (too) difficult to conduct, became one of the central ideas of cybernetics. W. Ross Ashby's (1956) *An Introduction to Cybernetics,* where the ideas and concepts of cybernetics were expounded extensively, contains a section on isomorphic machines and one on homomorphic machines. Two machines are isomorphic if one can be made identical to the other by relabeling. This relabeling can have various degrees of complexity, depending on what is relabelled: states or variables. If one of the two machines is simpler than the other, the simpler one is called a homomorphism of the more complex one. Although Ashby discusses homomorphism in terms of machines, he also means to include mathematical systems and models. This is nicely illustrated in figure 1.3, which shows three systems that are considered to be isomorphic.

Isomorphic systems are, according to Ashby, important because most systems have "both difficult and easy patches in their properties":

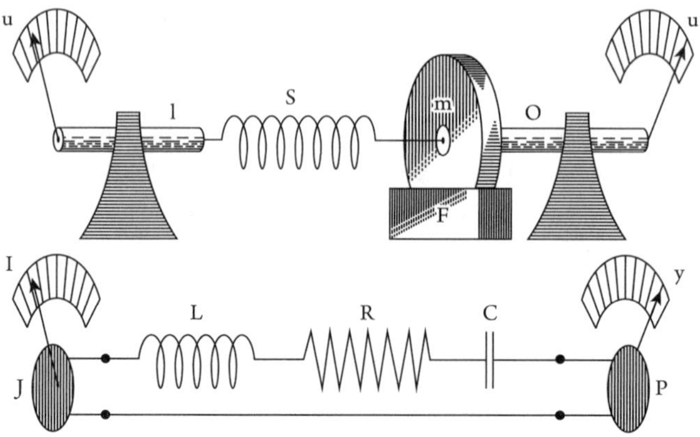

$$a\frac{d^2z}{dt^2}+b\frac{dz}{dt}+cz=w$$

Figure 1.3 Three isomorphic systems. Source: Ashby 1956, 95–96. Estate of W. Ross Ashby.

> When an experimenter comes to a difficult patch in the particular system he is investigating he may if an isomorphic form exists, find that the corresponding patch in the other form is much easier to understand or control or investigate. And experience has shown that the ability to change to an isomorphic form, though it does not give absolutely trustworthy evidence (for an isomorphism may hold only over a certain range), is nevertheless a most useful and practical help to the experimenter. In science it is used ubiquitously. (Ashby 1956, p. 97)

If interaction on a specific system is impossible, investigating an isomorphic, or homomorphic, system is supposed to provide the same level of understanding.

In the 1950s, the practice of modeling arose in several field sciences where experimentation was difficult or even impossible. The world of the model became the substitute world where "imaginative" experiments could be conducted.[17]

Besides this broadly shared notion of the "symmetrical roles" of models and experiments in the 1950s, there was also the shared idea about what kinds of experiments are specifically possible in social science, by contrast with natural science. The distinction that Morgenstern made between the first kind and the second kind of experiments is closely related to a distinction the philosopher Carl Hempel made in the same period. In a paper on the differences between methods in the natural and social sciences, Hempel (1952) distinguishes between two kinds of "imaginary experiments": the "intuitive" and the "theoretical." An imaginary experiment is aimed at anticipating the outcome of an experimental procedure that is just imagined. Anticipation is guided by past experience concerning particular phenomena and their regularities, and occasionally by belief in certain general principles that are accepted as if they were a priori truths. An imaginary experiment is called "intuitive" when

> the assumptions and data underlying the prediction are not made explicit and indeed may not even enter into the conscious process of anticipation at all: past experience and the—possibly unconscious—belief in certain general principles function here as suggestive guides for imaginative anticipation rather than as a theoretical basis for systematic prediction. (Hempel 1952, p. 76)

In contrast to the intuitive experiment, a "theoretical kind of imaginary experiment"

> presupposes a set of explicitly stated general principles—such as laws of nature—and it anticipates the outcome of the experiment by strict deduction from those principles in combination with suitable boundary conditions representing the relevant aspects of the experimental situation. (Hempel 1952, p. 76)

This experiment can be performed by a computer, similar to Morgenstern's experiments of the first kind, see above.

This distinction between "theoretical" and "intuitive" is subsequently used to distinguish between idealizations in economics and those in the natural sciences. According to Hempel the idealizations in economics are intuitive:

> the corresponding "postulates" are not deduced, as special cases, from a broader theory which covers also the nonrational and noneconomic factors affecting human conduct. No suitable more general theory is available at present, and thus there is no theoretical basis for an appraisal of the idealization involved in applying the economic constructs to concrete situations. (Hempel 1952, p. 82)

It should be noted that Hempel did not explicitly account for who should run the intuitive imaginary experiment, or who should have the appropriate economic intuitions. Morgenstern, however, did: it should be an economist, well informed by theory. But as economic theory is not exact, therefore it should be to "a very high extent in addition related to non-constructively obtained material, such as personal experience" (Morgenstern 1963a, p. 89).

In the 1950s, imaginary experiments were considered an appropriate alternative for laboratory experiments, a view that was strengthened by the contemporaneous development of the computer. But an experiment run on a computer—today called a simulation—requires "explicitly stated general principles," for example, laws of nature. In economics, there are not many of these principles, if there are any. Economics is an inexact science (see also chapter 4), which makes deductions such as predictions unreliable.

Olaf Helmer, like Hempel a student of Hans Reichenbach,[18] had developed not only rather similar ideas about the usage of statistical models as experiments but also about the incompleteness of models and theories in the social sciences. In a paper "On the Epistemology of the Inexact Sciences," which first appeared as a Rand working paper (1958) and was published the year after in *Management Science*, he, together with Nicholas Rescher, discussed what kind of knowledge was needed to complete a statistical model. The Rand paper opens with a long quotation from Alfred Marshall ([1890] 1920, p. 32) stating that the "laws of economics" are not "simple and exact" but "inexact and faulty" (see also chapter 3). Helmer saw these inexact laws, which he called "quasi-laws," as generalizations, less-than-universal principles, that are neither fully nor even explicitly articulated or even articulable. They require a different epistemology and methodology, namely, the systematic employment of expert judgment: "background information, which frequently may be intuitive in character and have the form of a vague recognition of underlying regularities, such as analogies, correlations, or other conformities" (Helmer and Rescher 1958, p. 30). For "quasi-laws,"

> statistical information matters less than knowledge of regularities in the behavior of people or in the character of institutions, such as traditions and customary practices, fashions and mores, national attitudes and climates of opinion, institutional rules and regulations, group aspirations, and so on. (Helmer and Recher 1958, p. 30)

For this kind of nonexplicit background knowledge we need experts and expertise. The expert has

> at his ready disposal a large store of (mostly inarticulated) background knowledge and a refined sensitivity to its relevance, through the intuitive application of which he is often able to produce trustworthy personal probabilities regarding hypotheses in his area of expertness. (Helmer and Rescher 1958, p. 31)

Helmer was a pioneer in futurology and always advocated the involvement of experts in forecasting and prediction:

> The key to progress in this field has been the recognition that in dealing with the future, especially in "soft" areas such as social, political, and economic developments, we have no firm laws providing the kind of predictive power associated with the laws of physics but must rely largely on intuitive understanding and perceptiveness of expert in relevant areas. (Helmer 1983, p. 20)

1.5. Measurement of Field Phenomena

The account of measurement that will be developed (in chapter 2) is a model-based account. This account is closely related to the representational theory of measurement. The core of this theory is that measurement is a process of assigning numbers to attributes or features of the empirical world in such a way that the relevant qualitative empirical relations among these attributes or features are reflected in the numbers themselves as well as in important properties of the number system. In other words, measurement is conceived of as establishing a homomorphism between a numerical and an empirical structure.[19] This numerical structure is considered to be a "representation" of the empirical structure, also called a "model" of the empirical structure.

The representational theory of measurement was developed in psychology because the experimentation on some psychological phenomena is not feasible. By first modeling the phenomenon of interest and drawing inferences from this model, the researcher is aiming to achieve the same reliable results as from experimentation in a laboratory. This kind of experimentation, also called a simulation, became even more feasible, particularly for complex systems, through the invention of the computer. Instead of experiments in the real material world, one can run simulations on models.

The account of model-based measurement that will be developed in chapter 2 will, however, be more general than the representational theory of measurement. It will be shown that homomorphic mapping is not the only rule that leads to reliable measurements. Homomorphic mappings imply white-box models. They are causal-descriptive statements on how real systems actually operate in some aspects. But to arrive at reliable measurement results, gray-box models, and even black-box models may also be used; it is only that their validation will be different.[20] Because field phenomena involve in principle an infinite number of potential influences of which an (unknown) part is unknown and another part is not measurable, any representation will inevitably be incomplete, "inexact," however large and comprehensive the model will be. This incompleteness makes the measurements based on homomorphic representations unreliable, but for a gray-box model this requirement of completeness is not needed to arrive at reliable measurement.

For measurement, considerations about reliability are closely related to considerations about objectivity. The reliability of measurement can be assessed objectively, because the employed measuring instruments can be validated as autonomous instruments, that is, can be investigated independently of the specific measurand or their user. In social science, these measuring instruments are usually not physical objects but models that function as autonomous measuring instruments (see Morrison and Morgan 1999 and Boumans 2005). For that reason, in social science the reliability of measurement can be assessed objectively by validating the employed models. But in case of the measurement of field phenomena, the model that is being used is only one part of the assessment of reliability. Because of its inherent incompleteness, a measurement needs additional expert judgment, as we have seen in the discussion of Morgenstern's work. The reliability of the measurement depends, beside the reliability of the model, also on the reliability of the expert judgments. Although the judgment itself can be considered as subjective, this does not necessarily imply that the measurement becomes subjective because of the involvement of expert judgments. This book will propose that these judgments can also be investigated independently, by validating expert knowledge in similar ways as models are assessed.

1.6. Conclusions

This book will develop in the following chapters an account of measurement for field sciences. Such an account will be necessarily different from an account of measurement for a laboratory science, because the nature of field phenomena differs inherently from the nature of laboratory phenomena. Laboratory phenomena are investigated in a controlled environment by systematic intervention, which allows the facts about these phenomena to be reproducible. As a result the knowledge acquired by this kind of research can be rather exact. Although field phenomena appear in environments that are designed by institutions, these institutions are usually not "scientific," and these phenomena are not reproducible. Knowledge about these phenomena is acquired in a much more "passive" way; that is, one can often plan and design the amount of observations and even plan and design what aspect of the phenomena can be "monitored," but direct intervention on the phenomenon itself is usually not possible.

This lack of control and systematic intervention is, however, compensated in field science by mapping the phenomenon and its environment into a mathematical model that is the virtual laboratory of the field scientist. But representing a field phenomenon and its environment by a mathematical model does not make our knowledge more exact. To complete our knowledge about field phenomena, the intuitions of experts are needed.

The involvement of experts conflicts with the current (laboratory) standards of objectivity. For measurement in field science we therefore need to consider what kind of alternative standards are needed to create objectivity. This will be the main topic of chapter 6.

Notes

1. Quoted in Manuel 1974, p. 120.
2. Quoted in Hempel 1945, p. 17.
3. It should be noted here that the methodological tension between the laboratory and the field is often expressed in terms of a tension between the "hard" standards of natural science and the "soft" standards of social science. Whenever this tension is mentioned, a closer look will often reveal that it is the standards of the laboratory that are referred to as being "hard" and "rigorous."

4. For more details on this interaction, see Leonard 2010. I would like to thank Robert Leonard for pointing me to Helmer.
5. Morgenstern 1928 has never been translated into English, but Arthur Marget's extensive review article (1929) of this book summarizes it adequately.
6. The Committee of Statistical Experts was established in 1928 to coordinate the unification and standardization of international statistics. This committee met once a year between 1931 and 1939. The committee members were not official delegates of their countries, but acted in their personal capacity. In addition to these members chosen from individual countries, the committee included one member from the International Labour Office and one from the International Institute of Agriculture. The convention contained no limitation as to the number of committee members, but during most of the committee's existence the number was ten. There were no official representatives from the International Statistical Institute, although most committee members were also members of this institute. Of the methodological studies made by the committee, six were published in the League of Nations series Studies and Reports on Statistical Methods. For more details, see Nichols 1942.
7. It is the same problem as Galileo's ([1632] 1967) discussion of the problem of determining the position of the new star of 1572. See chapter 3 for a more detailed discussion of the treatments of errors.
8. There exist various translations, of which Abraham Pais's is my favorite: "Subtle is the Lord, but malicious He is not." Pais (1982, p. 114) provides the following story that goes with it: Oswald Veblen, a professor of mathematics at Princeton and a nephew of Thorstein Veblen, wrote in 1930 to Albert Einstein, asking his permission to have this statement chiseled in the stone frame of the fireplace in the common room of Fine Hall. In his reply to Veblen, Einstein consented and gave the following interpretation of his statement: "Die Natur verbirgt ihr Geheimnis durch die Erhabenheit ihres Wesens, aber nicht durch List," translated as "Nature hides its secret because of its essential loftiness, but not by means of ruse."
9. Bowman was president of the American Statistical Association in 1963.
10. Kuznets was president of the American Statistical Association in 1949.
11. This history of a mathematical theory of error is well exposited by Judy Klein (1997), L. E. Maistrov (1974), and Stephen Stigler (1986).
12. Parts of this memorandum have been reproduced in Helmer 1983.
13. To my knowledge, this is the first paper in which the principles of the representational theory of measurement are set out, though not yet under this name.
14. At the Santa Monica seminar, where Coombs, Raiffa, and Thrall (1954a, 1954b) discussed the "symmetrical roles" of experiments and mathematical models, Morgenstern presented his paper "Experiment and Computation in Economics" with a similar message.
15. Morgan (2005) has a rather similar account, though in her account both routes are not symmetrical in epistemological power.
16. Mathematisation and measurement are closely related. Several examples of this close relationship will be discussed in the subsequent chapters. This relationship is also emphasized by Brian Ellis (1966). The first sentence of the introduction to his *Basic Concepts of Measurement* reads: "Measurement is the link between mathematics and science" (p. 1).
17. See Morgan 2012, for a rather detailed account of "imaginative" experiments on the "virtual" phenomena of the economic worlds in models.
18. According to Rescher (2006), Hempel and Helmer constitute the "middle generation" of the Berlin school of logical empiricism because both were students of Reichenbach. For more biographical details and the interrelationships between Hempel and Helmer, see Rescher 2006.
19. The term derives from the Greek *omo*, "alike," and *morphosis*, "to form" or "to shape." It denotes that the assigned numerical relational system preserves the properties of the empirical relational structure.
20. Gray-box models will be defined and discussed in chapter 2: They are modular-designed representations, that is, assemblies of modules. Modules are black boxes with standard interface.

2

Measurement

In physical science a first essential step in the direction of learning any subject is to find principles of numerical reckoning and methods for practicably measuring some quality connected with it. I often say that when you can measure what you are speaking about, and express it in numbers, you know something about it; but when you cannot express it in numbers, your knowledge is of a meagre and unsatisfactory kind; it may be the beginning of knowledge, but you have scarcely, in your thoughts, advanced to the stage of *science*, whatever the matter may be.

—William Thomson (1889, pp. 73–74)

Until the phenomena of any branch of knowledge have been subjected to measurement and number, it cannot assume the status and dignity of a science.

—Francis Galton (1879, p. 149)[1]

Through measurement to knowledge.

—Heike Kamerlingh Onnes, 1882[2]

2.1. Introduction

Science is measurement.[3] According the standard view on science, usually called the "Received View," science is objective knowledge and measurement is the preeminent scientific method to obtain this kind of knowledge.[4] But as much as science is not entirely objective, neither is measurement. The consequences of this simple observation will be explored in this chapter.

To see what is at stake, we can look at a current debate in metrology on how to express the reliability of a measurement result.[5] In metrology, the traditional approach is to express this reliability in terms of "error." The traditional approach takes it for granted that a measurand can ultimately be described by as a single true value, but that instruments and measurements do not yield this value due to additive "errors." In current metrology, however, it is recognized that, when all of the known or suspected components of error have been evaluated and the appropriate corrections have been applied, there still remains an uncertainty about the correctness of the stated result, that is, a doubt about how well the result of the measurement represents the value of the quantity being measured (see JCGM 100 2008, p. viii). Therefore, in the new uncertainty approach, as it is called, the notion of error no longer plays a role; there is finally only one "uncertainty" of measurement. A basic premise of the uncertainty approach is that

> it is not possible to state how well the essentially unique true value of the measurand is known, but only how well it is believed to be known. Measurement uncertainty can therefore be described as a measure of how well one believes one knows the essentially unique true value of the measurand. This uncertainty reflects the incomplete knowledge of the measurand. (JCGM 104 2009, p. 3)

So, instead of evaluating measurement results in terms of errors, it is now preferred to assess the reliability of measurement in terms of uncertainty.

To evaluate uncertainty of measurement results, it is recommended in metrology to use two different ways of evaluating uncertainty components, a Type A evaluation and a Type B evaluation:

Type A evaluation is the "method of evaluation of uncertainty by the statistical analysis of series of observations" (JCGM 100 2008, p. 3).

Type B evaluation is the "method of evaluation of uncertainty by means other than the statistical analysis of series of observations" (JCGM 100 2008, p. 3).

Type A evaluation can be objectively established as soon as a metric is chosen, as it is a quantitative concept. Type B evaluation, however, is not based on a series of observations. It is considered to be a "scientific judgement" based on professional skill "that can be learned with practice" (JCGM 100 2008, p. 12) depending on qualitative and subjective knowledge of the measurand and "experience with or general knowledge of the behavior and properties of relevant materials and instruments (JCGM 100 2008, p. 11). Type B uncertainty cannot therefore be evaluated in the same objective way as Type A uncertainty.

That objective standards are not enough for evaluating measurement results is admitted in one of the key publications in metrology, the *Guide to the Expression of Uncertainty in Measurement*:[6]

> Although this *Guide* provides a framework for assessing uncertainty, it cannot substitute for critical thinking, intellectual honesty and professional skill. The evaluation of uncertainty is neither a routine task nor a purely mathematical one; it depends on detailed knowledge of the nature of the measurand and of the measurement. The quality and utility of the

> uncertainty quoted for the result of a measurement therefore ultimately depend on the understanding, critical analysis, and integrity of those who contribute to the assignment of its value. (JCGM 100 2008, p. 8)

Type B evaluation of uncertainty requires a more "personal" judgment, hence an emphasis on the personal responsibility for a scientific attitude on the part of the individual researcher, indicated by scientific values such as "honesty" and "integrity." Similar appeals to personal scientific attitudes are made in other fields where the necessity of this kind of evaluation is acknowledged as part of measurement. Take, for example, a document of the CPB Netherlands Bureau for Economic Policy Analysis in which it is explained how this bureau arrives at its forecasts of economic developments in the Netherlands. Due to the uncertainty of the (economic) future, the document emphasizes that "Common sense and a critical attitude with respect to the model, which by definition can never be more than a simplified representation of reality, are essential" (de Jong, Roscam Abbing, and Verbruggen 2010, p. 15; translated by the author).

Measurement involves a combination of mechanical objectivity and "trained judgement" (Daston and Galison 2007). Because objectivity involves personal judgment of an expert, for the assessment of the reliability of a measurement, scientific values as critical thinking, intellectual honesty, and professional skill have to be taken into account.

With respect to "critical thinking" and "intellectual honesty," I here notice only that they are part of reliability assessments, and will not discuss how to assess them. The assessment of professional skill, however, will be considered in more detail in this book. Chapter 6 discusses strategies to quantify expertness, which will be used to make measurement more objective.

Besides the implicit connotation of objectivity, the claim "Science is measurement" also refers to a specific image of science: If measurement in a particular discipline is not feasible, then that particular discipline cannot be considered a science. Because the traditional concept of measurement is based on measurement in physics, the scientific content of field science, as well as social science, was and still is contested. It is for that reason that alternative theories of measurement were developed, of which the representational theory of measurement is the most prominent one.

The traditional, "classical," concept of measurement is that measurement is "the discovery or estimation of the ratio of some magnitude of a quantitative attribute to a unit (a unit being, in principle, any magnitude of the same quantitative attribute)" (Michell 1999, p. 14). This concept of measurement can be found, for example, in James Clerk Maxwell's *Treatise on Electricity and Magnetism*, in the "Preliminary on the Measurement of Quantities" (1873, p. 1):

> Every expression of a Quantity consists of two factors or components. One of these is the name of a certain known quantity of the same kind as the quantity to be expressed, which is taken as a standard of reference. The

> other component is the number of times the standard is to be taken in order to make up the required quantity. The standard quantity is technically called the Unit, and the number is called the Numerical Value of the quantity.

The most influential twentieth-century version of this classical concept of measurement was Norman R. Campbell's account set out in his *Physics: The Elements* (1920). The problem with this classical concept is that it is most adequate for measurement in physics but excludes most if not all measurements in the social sciences.

The representational theory of measurement was developed to account for nonphysical phenomena. The core of this theory is that measurement is a process of assigning numbers to attributes or characteristics of an empirical phenomenon (measurand) in such a way that the relevant qualitative empirical relations among these attributes or characteristics are reflected in the numbers themselves as well as in important properties of a numerical relational structure. This numerical relational structure is a representation of the empirical relational structure of the measurand. This representation is also called a model; therefore, the representational theory of measurement is sometimes called the *model theory of measurement*.

As Michael Heidelberger (1994) observes, a first glimpse of the origins of the representational theory of measurement can be traced to Maxwell's method of using formal analogies.[7] In Maxwell's article "On Faraday's Lines of Force" ([1856] 1965) when discussing his method of using analogies, Maxwell adopts the "representational view" en passant: "Thus all the mathematical sciences are founded on relations between physical laws and laws of numbers, so that the aim of exact science is to reduce the problems of nature to the determination of quantities by operations with numbers" (p. 156). Hermann von Helmholtz took up Maxwell's view and continued to think in the same direction. Michell (1993, 1999) and Savage and Ehrlich (1992) take Helmholtz's 1887 article "Zählen und Messen, erkenntnis-theoretisch betrachtet" (published in translation as "An Epistemological Analysis of Counting and Measurement," in Kahl 1971) as the starting point of the development of the representational theory.

As a result of this "representational" view on measurement, the fundamental problem of measurement is the formulation of the criteria needed for assessing whether a model does represent an empirical system and in what way. This is called the "representation problem." Generally there are two different "foundational" approaches to deal with this representation problem: an axiomatic approach (discussed in section 2.2) and an empirical approach.

The axiomatic approach was developed in mathematical psychology as a response to Campbell's concept of measurement that disqualified measurement in psychology. According to Campbell's account, meaningful measurement is based on experimental procedures to combine physical objects. These procedures are, however, not possible for the kind of entities (e.g., sensory events) measured in psychology. To circumvent this impossibility, the axiomatic approach is in no sense based on experimental procedures, but consists of formal restrictions on the model assumptions.

Outside the disciplinary boundaries of mathematical psychology, the representational theory of measurement was accepted as providing logical foundations for measurement, but was criticized for lacking empirical foundations. In measurement science these foundations were developed by defining empirical criteria to validate a model (discussed in section 2.3). These criteria, however, are not easily transferred to a field science. The application of these criteria to measurement of field phenomena runs the risk, as will be discussed in section 2.4, that model validation becomes too subjective. Therefore, section 2.4 will propose a different kind of model structure to enable more objective validation procedures. In summary, this chapter aims to extend the representational theory of measurement in such a way that this more general framework accounts for measurement of social field phenomena.

2.2. Axiomatic Theory of Measurement

The axiomatic version of the representational theory is the account of measurement that is presented in the three-volume survey *Foundations of Measurement*, edited by David H. Krantz, R. Duncan Luce, Patrick Suppes, and Amos Tversky (1971, 1989, 1990). A quick glance at these volumes, however, would not immediately give the impression that they contain an account of different kinds of measurement, but rather a theory of various functional-theoretic algebras. The reason for this highly formalist presentation is that the "foundations of measurement" are established by axiomatization. Therefore, this formal version of the representational theory of measurement will be labeled as the *axiomatic theory of measurement*. This theory focuses on the properties of the numerical assignments, rather than on the experimental or instrumental procedures for making these assignments. According to the axiomatic theory of measurement the foundations of measurement are established by the explication and axiomatization of the properties of these numerical assignments.

The axiomatic theory originated as a response to a discussion within the British Association for the Advancement of Science in the 1930s on the meaning of "measurement" or "quantitative estimate," namely, whether or not these terms should be used "in any sense other than that in which they are used in physics" (Ferguson et al. 1940, p. 332). The original problem was whether sensation intensities are measurable or not, but this grew into the more general problem whether quantitative estimation of nonphysical entities could rightly be called measurement.

To discuss this problem a committee was appointed by sections A (mathematical and physical sciences) and J (psychology) of the British Association for the

Advancement of Science in 1932 (and reappointed each year since) to consider and report on the "possibility of quantitative estimates of sensory events" (Ferguson et al. 1940, p. 331). The committee was chaired by Allan Ferguson and had Campbell as one of its members.[8] The committee, however, did not come to an agreement in the final report, and therefore individual members—"who cared to do so"— were invitedto express their views. These views appeared as appendices to the report.

Appendix 2 was on measurement, and consisted of two statements, of which Campbell's was the longest. (The other was only a very short note on the description of estimated magnitudes.) Campbell as author of *Physics: The Elements* (1920) and *Account of the Principles of Measurement and Calculation* (1928) was at that time the authority on measurement. Campbell defined measurement "in its widest sense" as "the assignment of numerals to things so as to represent facts or conventions about them" (Campbell 1940, p. 340).[9] However liberal this definition may look, he restricted its scope by stating that in physics only two kinds of measurements can be distinguished: "direct" and "indirect." It is, however, more conventional in physics, including Campbell's own work, to use the labels "fundamental" and "derived." Therefore these latter labels will hereafter be used.

For "fundamental" measurement Campbell imposed the following condition: "when a numeral has been assigned to one member of the group, the numeral to be assigned to any other member is or can be determined by facts within a limited range of 'experimental error,' arising from the nature of the facts that determine the assignment" (Campbell 1940, p. 340). According to Campbell there is only one way to fulfil this condition, namely the "rule" that

> the numeral to be assigned to any thing X in respect to any property is that which represents the number of standard things or "units," all equal in respect of the property, that have to be combined together in order to produce a thing equal to X in respect of the property. (Campbell 1940, p. 340)

In other words, the rule has to meet the restrictions of the classical concept of measurement. The only properties measurable by means of this rule are those that are additive, that is, "which are such that, given two things A and B having the property, it is possible to produce by a precisely determined operation (combination) a thing C which is greater in respect of the property than either A or B" (p. 340). "Typical" additive properties with their combinations are the following:

> Length—placing end to end in a straight line
> Mass—connecting so as to form a single rigid body
> Electrical resistance—connected in series (Campbell 1940, p. 340)

The measuring rule here is thus a procedure of combining things (see also Ellis 1966, p. 52).

Nonadditive properties are measured by an "indirect process," of which the simplest form is the following:

> To each thing to be measured is assigned by the direct process two numerals, x, y, representing it in respect of two additive properties. The fact is established that the numerals representing some single-valued function, $f(x, y)$, have the same order as the non-additive property to be measured. The numeral $f(x, y)$ is then assigned to represent the property. (Campbell 1940, pp. 340–341).

The example Campbell gave is density = mass / volume.[10] He emphasized that derived measurements "require the establishment by experiment of a law relating directly measured properties" (p. 341).

Thus, according to Campbell, all measurement requires the establishment of facts either by the procedure of additivity or by experiment that shows that the assumed law between previously measured properties is true. He left no ambiguity regarding what he thought about so-called measurement in psychology:

> True measurement is relegated nowadays to instrument makers and standardizing laboratories, who make standards or calibrated instruments which permit other things to be measured by mere judgment of equality. (Campbell 1940, p. 341)

This definition of measurement was meant to be exclusive. But, as Campbell admitted, the definition was perhaps too exclusive. What if a law is being used to measure a property that is far outside the range over which this law is verified by experimental result? Then two possibilities arise. One is that another process of measuring is used, based on relations other than the previous law over a range where the property is supposed to exist. The other possibility is that no process can be found for measuring this property. If so, the measurement of that property rests "on purely arbitrary conventions that have no factual basis" (p. 342), and so is not "true measurement."

> But the possibility of such an attitude is proof that physicists have very definite criteria of measurement, and that measurement is possible only in virtue of facts that have to be proved and not assumed. (Campbell 1940, p. 342)

The committee was of the opinion that its position would be made clearer if the points expressed in the report were illustrated by reference to a concrete example of a sensory scale. They chose for this purpose the Sone scale of loudness as expounded in *Hearing: Its Psychology and Physiology*, by Stanley Smith Stevens and Hallowell Davis (1938). As a response to this report, in particular the way measurement was defined by Campbell such that it excluded his scale of loudness, Stevens (1946) designed a classification of scales of measurement to include measurement in psychology.

To arrive at this classification, he first paraphrased Campbell's definition of measurement: "measurement, in the broadest sense, is defined as the assignment of numerals to objects or events according to rules" (Stevens 1946, p. 677). Different rules lead to different kinds of scales and so to different kinds of measurement. The problem is then making explicit

> a) the various rules for the assignment of numerals
> b) the mathematical properties (or group structure) of the resulting scales, and
> c) the statistical operations applicable to measurements made with each type of scale. (Stevens 1946, p. 677)

A scale is then "a certain isomorphism between what we can *do* with the aspects of objects and the properties of the numeral series" (p. 677, emphasis added). This numeral series can be used "as a *model* to represent aspects of the empirical world" (p. 677).

Although in the subsequent literature the scales of measurement came to be classified according to their "mathematical group structure" (see Ellis 1966, p. 58), Stevens, like Campbell, still aimed at a classification based on "empirical operations" ("what we can *do*"):

> The type of scale achieved depends upon the character of the basic empirical operations performed. These operations are limited ordinarily by the nature of the thing being scaled and by our choice of procedures, but, once selected, the operations determine that there will eventuate one or another of the scales listed in Table 1. (Stevens 1946, pp. 677–678)

But in the table (see table 2.1) these "basic empirical operations" looked very much like mathematical procedures; there was no mentioning of, for example, the procedure of additivity. The column of the "mathematical group structure" lists the mathematical transformations that leave the scale-form invariant.

Stevens's concept of measurement can be considered an intermediate phase in the development from Campbell's account requiring experimental procedures to the abstract axiomatic theory. While Campbell's definition of derived measurement is still based on an empirically verified law, Stevens saw derived measurements as "mathematical functions" of fundamental measurements. Moreover, the problem

Table 2.1 **Stevens's table of scales**

Scale	*Basic empirical operations*	*Mathematical group structure*	*Permissible statistics (invariantive)*
Nominal	Determination of equality	*Permutation group* $x' = f(x)$ $f(x)$ means any one-to-one substitution	Number of cases Mode Contingency correlation
Ordinal	Determination of greater or less	*Isotonic group* $x' = f(x)$ $f(x)$ means any monotonic increasing function	Median Percentiles
Interval	Determination of equality of intervals or differences	*General linear group* $x' = ax + b$	Mean Standard deviation Rank-order correlation Product-moment correlation
Ratio	Determination of equality of ratios	*Similarity group* $x' = ax$	Coefficient of variation

Source: Stevens 1946, table 1, p. 678. Reprinted with permission from AAAS.

"as to what measurement is" was reduced by him to the question "what the rules of assignment are":

> If we can point to a consistent set of rules, we are obviously concerned with measurement of some sort, and we can then proceed to the more interesting question as to the kind of measurement it is. In most cases a formulation of the rules of assignment discloses directly the kind of measurement and hence the kind of scale involved. If there remains any ambiguity, we may seek the final and definitive answer in the mathematical group structure of the scale form. (Stevens 1946, p. 680)

In the development of measurement theory, Stevens's approach was a step away from experimental procedures to a more mathematical concept of measurement, but not (yet) completely: "Measurement is never better than the empirical operations by which it is carried out" (p. 680). That Stevens's account of measurement was an intermediate position between empirical foundations and axiomatic foundations became apparent at the 1956 meetings of the American Association for the Advancement of Science, which included a symposium on measurement.[11] The step away from a purely empirical foundation was made by adding "any rule" to his original 1946 definition of measurement (previously quoted):

> Measurement [is] the assignment of numerals to objects or events according to rule—*any rule*. Of course, the fact that numerals can be assigned under different rules leads to different kinds of scales and different kinds of measurement, not all of equal power and usefulness. Nevertheless, provided a consistent rule is followed, some form of measurement is achieved. (Stevens 1956, p. 19; emphasis added)

But he left no ambiguity about his view that measurement still needs an empirical foundation. He emphasized that notwithstanding mathematics that had become "a game of signs and rules having no necessary reference to empirical objects or events" (p. 20), measurement "must remain anchored here below, for it deals with empirical matters" (p. 19).

Stevens's account of measurement is usually aligned with his view on science, which he labeled "operationism" and consists of the following "operational principles":[12]

1. "Science, as we find it, is a set of empirical propositions agreed upon by members of society. This agreement may be always in a state of flux, but persistent disagreement leads eventually to rejection." (Stevens 1939, p. 227)
2. "Only those propositions based upon operations which are public and repeatable are admitted to the body of science." (p. 227)
3. "A term denotes something only when there are concrete criteria for its applicability; and a proposition has empirical meaning only when the criteria of its truth or falsity consist of concrete operations which can be performed upon demand." (p. 228)

Measurement is based on operations, but these operations are not restricted to additive combinations or relations based on laws. Any operation that is "public," that is, "agreed upon by members of society" and "repeatable," is in principle appropriate for measurement. Which operation will be chosen depends on its applicability with respect to the measurand.

Coombs, Raiffa, and Thrall (1954a, 1954b) constructed the next step away from Campbell's empirical procedures by considering Stevens's "Basic Empirical Operations" (see table 2.1) as axioms.[13] The different scales were now supposed to be defined by axioms. Because these axioms were disconnected from any concrete empirical procedure, Coombs, Raiffa, and Thrall could come up with a broader set of axioms, and so they arrived at a more extensive system of scales. For example, a "weak order" scale was defined by the axiom that for every pair a, b either $a \leq b$ or $b \leq a$ (p. 142). Nevertheless, these axioms were still supposed to be "satisfied by that segment of the real world which is mapped into it"; otherwise the measurement would have no meaning.

To frame measurement in terms of axioms was inspired by John von Neumann and Oskar Morgenstern's *Theory of Games* (1953).[14] For the measurement of utility, von Neumann and Morgenstern provided an "axiomatic treatment." The requirement for such a treatment was to find a "correspondence" between utilities and numbers that carries the relation $u > v$ and the operation $\alpha u + (1-\alpha)v$ $(0 < \alpha < 1)$ for utilities into the "synonymous concepts" for numbers (p. 24). Both relation and operation were considered to be "natural" in the domain of utility. So the numerical assignment ("correspondence") $u \rightarrow \rho = v(u)$ had to be order-preserving, that is, $u > v$ implies $\mathrm{v}(u) > \mathrm{v}(v)$, and linear, that is, $\mathrm{v}(\alpha u + (1-\alpha)v) = \alpha\mathrm{v}(u) + (1-\alpha)\mathrm{v}(v)$. Subsequently, von Neumann and Morgenstern showed that the "numerical valuation of utilities" is "determined up to a linear transformation" (p. 25). In other words, without labeling it as such, they showed that the appropriate scale for the measurement of utilities is the interval scale (see table 2.1).

In order to ensure the existence of such a homomorphic ("order-preserving") numerical assignment, they "postulated" the following set of "axioms": Consider a "system" U of entities $u, v, w, \ldots$

1. $u > v$ is a complete ordering U.
   This means that
   a. For any two u and v one and only one of the three relations hold: $u = v, u > v, u < v$.
   b. $u > v, v > w$ imply $u > w$.
2. Ordering and combining.
   a. $u < v$ implies that $u < \alpha u + (1-\alpha)v$.
   b. $u > v$ implies that $u > \alpha u + (1-\alpha)v$.
   c. $u < w < v$ implies the existence of an α with
   $$\alpha u + (1-\alpha)v < w.$$
   d. $u > w > v$ implies the existence of an α with
   $$\alpha u + (1-\alpha)v > w.$$
3. Algebra of combining
   a. $\alpha u + (1-\alpha)v = (1-\alpha)v + \alpha u$.
   b. $\alpha(\beta u + (1-\beta)v) + (1-\alpha)v = \gamma u + (1-\gamma)v$, where $\gamma = \alpha\beta$.
   (von Neumann and Morgenstern 1953, p. 26)

This set of axioms became the template for the subsequent developments in the axiomatic approach to measurement.

"Basic Measurement Theory" by Suppes and Joseph L. Zinnes (1963) can be considered the first version of the axiomatic theory of measurement. This "basic theory" was published as the first chapter in volume 1 of the three-volume *Handbook of Mathematical Psychology*, edited by R. Duncan Luce, Robert R. Bush, and Eugene Galanter. These volumes were instrumental in the definition, establishment, and further development of the new field of mathematical psychology. Because quantification of behavior was considered to be fundamental in this new field, the theory of measurement was a central topic ("one of the gods" [p. 3]) in mathematical psychology; hence it was given a prominent position as the first chapter of the *Handbook*.

The theory was set up to answer two "fundamental problems." The first problem, called the "representation problem," was the "justification of the assignment of numbers to objects or phenomena" (p. 2). The second problem, the "uniqueness problem," concerned "the specification of the degree to which this assignment is unique" (p. 4).

The representation problem was more "completely" stated as follows:

> Characterize the formal properties of the empirical operations and relations used in the procedure and show that they are isomorphic to appropriately chosen numerical operations and relations. (Suppes and Zinnes 1963, p. 4)

To arrive at a more formal expression of the representation problem, they used Tarski's (1954) notion of a "relational system":

> A relational system is a finite sequence of the form $\mathsf{A} = < A, R_1, \ldots, R_n >$, where A is a nonempty set of elements called the *domain* of the relational system A, and $R_1, \ldots, R_n$ are relations on A. (Suppes and Zinnes 1963, p. 5)

This definition of a relational system was subsequently used to define an "isomorphic image":

> Let $\mathsf{A} = < A, R_1, \ldots, R_n >$ and $\mathsf{B} = < B, S_1, \ldots, S_n >$ be similar relational systems. Then B is an *isomorphic* image of A if there is a one-one function f from A to B such that, for each $i = 1, \ldots, n$ and for each sequence $< a_1, \ldots, a_{m_i} >$ of elements of A, $R_i(a_1, \ldots, a_{m_i})$ if and only if $S_i(f(a_1), \ldots, f(a_{m_i}))$. (Suppes and Zinnes 1963, p. 6)

Because objects may have, for example, the same length, they suggested that for these cases one should weaken the requirement of f to be "one-one" and instead to speak of B as the *homomorphic* image of A.

Thereupon, a distinction was made between a "numerical relational system" and an "empirical relational system." A numerical system is a relational system whose domain is a set of real numbers, and an empirical system is a relational system whose domain is "a set of identifiable entities, such as weights, persons, attitude statements, or sounds" (p. 7).

So the more formal statement of the "first fundamental problem of measurement" was cast as

> the problem of showing that any empirical relational system that purports to measure (by a simple number) a given property of the elements in the domain of the system is isomorphic (or possibly homomorphic) to an appropriately chosen numerical relational system. (Suppes and Zinnes 1963, p. 7)

The second problem, uniqueness, was formulated in the following way: "determine the scale type of measurements resulting from the procedure" (p. 10). The reason to call it the "uniqueness problem" was that

> from a mathematical standpoint the determination of the scale type of measurements arising from a given system of empirical relations is the determination of the way in which any two numerical systems are related when they use the same numerical relations and are homomorphic to the given empirical system. (Suppes and Zinnes 1963, p. 10)

In other words, the determination of the appropriate scale was no longer decided by the "basic empirical operation," column 2 in Stevens's table (table 2.1), but by the "mathematical group structure," column 3 in that table.

To see this more precisely, consider first their formal definition of "scale":[15]

> Let A be an empirical relation system, let $\mathfrak{R}$ be a full numerical relation system, and let f be a function that maps A homomorphically onto a subsystem of $\mathfrak{R}$. . . . We say then that the ordered triple $< \mathsf{A}, \mathfrak{R}, f >$ is a *scale*. (Suppes and Zinnes 1963, p. 11)

The determination of a scale, as resulting from the "uniqueness theorem," follows now from the "admissible transformation" ϕ: Let $< \mathsf{A}, \mathfrak{R}, f >$ be a scale and g be any function having the property that $< \mathsf{A}, \mathfrak{R}, g >$ is also a scale. Then the kind of transformation ϕ for which $g = \phi \circ f$ determines the type of scale.[16] These transitions are the "structures" f in table 2.1. For example, the linear transformation $\phi(x) = ax + b$ determines the interval scale.

To point out the difference between Stevens's original, more empirical view of measurement—"some writers of measurement theory appear to define scales in terms of the existence of certain empirical operations" (p. 15)—and their own, Suppes and Zinnes stated quite clearly that in their formulation of the scale,

> no mention is made of the kinds of "direct" observations or empirical relations that exist (in the empirical relational system). Scale type is defined entirely in terms of the class of numerical assignments which map a given empirical system homomorphically onto a subsystem of the same full numerical system. If in a given instance these numerical assignments are related by a linear transformation, then we have an interval scale. Precisely what empirical operations are involved in the empirical system is of no consequence. . . . For example, instead of asking how we know certain intervals are "really" equal, we ask if all the admissible numerical assignments are related by linear transformation. (Suppes and Zinnes 1963, p. 15)

As in Stevens's account, the definition of derived measurement was not connected to empirical relations: "derived measurement does not depend on an empirical relational system directly but on other numerical assignments" (p. 17). So Suppes and Zinnes's "basic theory" was a formal theory with a very loose, if any, connection to empirical operations.

But it should be noted here that the "basic measurement theory" of Suppes and Zinnes was still not the highly formalistic axiomatic theory of measurement that it became in the three-volume survey *Foundations of Measurement* (1971, 1989, 1990). In addition to fundamental and derived measurements, Suppes and Zinnes also discussed "pointer measurement": "a numerical assignment (either fundamental or derived) based on the direct readings of same validated instrument" (p. 20). This kind of measurement is nowhere discussed in the *Foundations.*

Notwithstanding that the construction of a measuring instrument is based on "some established law" or theory, the discussion of "pointer measurement" was only of theoretical interest and related to the two "fundamental problems of measurement": "the problem of justifying the numeral assignment, in this case the readings of the instrument, and that of specifying the uniqueness of the numerical assignment" (p. 21). The representation problem was solved by how the instrument was considered to be validated: "if it has been shown to yield numerical values that correspond to those of some fundamental or derived numerical assignment . . . under 'standard' conditions" (p. 20). This is calibration. Recall that according to Campbell "true measurement" was equivalent to measurement with calibrated instruments. It seems that in 1963, Campell's physicalist view on measurement still could not be fully ignored in psychology.

The uniqueness of the readings, however, was determined by the uniqueness of the corresponding fundamental or derived numerical assignment, and "not, as might appear, by the method of calibrating the pointer instrument" (p. 22). So the scale type derives directly from the scale type of the corresponding numerical assignment.

The development sketched thus far culminated into the three-volume *Foundations of Measurement* (1971, 1989, 1990). Foundations meant mathematical foundations, more precisely, axiomatic foundations. The axioms are considered to be the "few explicit properties from which all others are deduced" (Krantz et al. 1971, p. 7). So the axioms are the properties of objects or events, and not the "explication and systematization of the assumptions required by particularly interesting procedures of measurement" (p. 7), and certainly do not represent Campbell's empirical "concatenation operations."

> The absence of appropriate, empirically defined, concatenation operations in psychology has even led some serious students of measurement to conclude that fundamental measurement is not possible there in the same sense that it is possible in physics. . . . Many examples in this book show that Campbell's viewpoint is untenable. (Krantz et al. 1971, p. 7)

This "concatenation operation" is Campbell's procedure of additivity, without which there is, according to Campbell, no fundamental measurement:[17]

> A major difficulty in most attempts to apply the theory of extensive measurement to nonphysical attributes such as utility, intelligence, or loudness is the lack of an adequate interpretation for the concatenation operation. Indeed, this lack has led some authors, such as Campbell (1920) . . . , to conclude that fundamental measurement is impossible in the social sciences. (Krantz et al. 1971, p. 123)

This "impossibility" was simply bypassed by stating that the focus of the foundations of measurement is on the properties of the numerical assignment, rather than on the procedure for making the assignment (p. 8). Krantz and coauthors simply *assumed* that these properties are empirical properties, some of which could be formulated as axioms:

> A set of axioms leading to representation and uniqueness theorems of fundamental measurement may be regarded as a set of qualitative (that is, nonnumerical) empirical laws. (Krantz et al. 1971, p. 13)

The axiomatic theory of measurement was developed to show that measurement is possible in the social sciences, by shifting the focus from the representational problem—an empirical problem—to the uniqueness problem—a mathematical problem. In doing so it became a theory that lacks any empirical foundation, or as another proponent of the axiomatic theory of measurement put it succinctly:

> We usually do not attempt to describe an actual process that is undergone as techniques of measurement are developed. We are not interested in a measuring apparatus and in the interaction between the apparatus and

the objects being measured. Rather, we attempt to describe how to put measurement on a firm, well-defined foundation. (Roberts 1979, p. 2)

And this foundation was nothing else than an axiomatic foundation.

2.3. Metrological Representational Theory of Measurement

> It is imperative to notice that whenever we apply a definition to nature we must wait to see if it will correspond to it. With the exception of pure mathematics we can create our concepts at will, even in geometry and still more in physics, but we must always investigate whether and how reality corresponds to these concepts.
>
> —Ernst Mach, "Critique of the Concept of Temperature" ([1896] 1966, p. 185)[18]

While the axiomatic theory of measurement was mainly developed and used within mathematical psychology, its more general version, the representational theory of measurement, received interest from instrument and control engineering as a possible way to give measurement in engineering "epistemological and logical foundations." Actually it was Ludwik Finkelstein (1975) who introduced the representational theory of measurement to measurement science, to bridge "the gap between the abstract philosopher's approach and that of the pragmatic instrument designer" (Finkelstein 1982, p. 1). His version of the representational theory of measurement was developed from Ellis (1966), Pfanzagl (1968), and Krantz, Luce, Suppes, and Tversky (1971). "Along the lines developed by Suppes and Zinnes," he defined measurement as follows:

> Take a well-defined, non-empty class of extra-mathematical entities Q (Fig 1). Let there exist upon that class a set of empirical relations $\Re = \{R_1, \ldots, R_n\}$. Let us further consider a set of numbers N (in general a subset of the set of real numbers Re) and let there be defined on that set a set of numerical relations $\wp = \{P_1, \ldots, P_n\}$. Let there exist a mapping M with domain Q and a range in $N, M : Q \to N$ which is a homomorphism of the empirical relational system $< Q, \Re >$ and the numerical relational system $< N, \wp >$. Then the triplet $S =< Q, N, M >$ constitutes a scale of measurement of Q. (Finkelstein 1975, p. 105)

Finkelstein illustrated this complex definition by a "diagrammatic representation of the set-theoretical definition of measurement" (see figure 2.1).

In this definition of measurement, the representation problem is accounted for by explicitly mentioning that the mapping is a homomorphism. But the second fundamental problem of the axiomatic theory of measurement, the uniqueness problem, played only a minor role. In one of the last sections of Finkelstein's

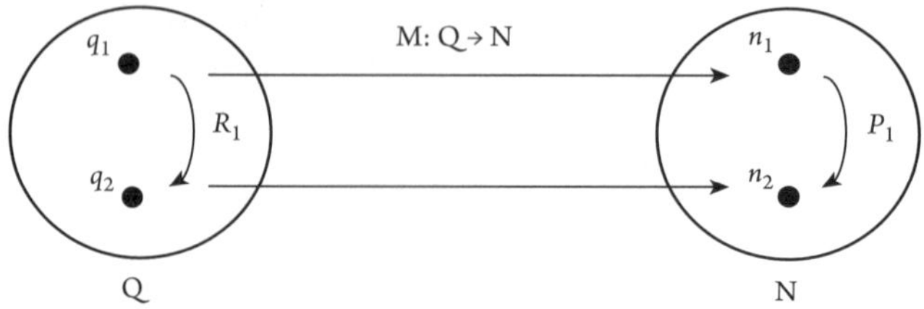

Figure 2.1 Representational Theory of Measurement. Source: Adapted from Finkelstein 1975, p. 105, with permission of *Measurement + Control.*

1975 article "Fundamental Concepts of Measurement," scale transformations and the uniqueness of a mapping were only discussed very briefly. Instead of emphasizing the need for solving uniqueness problems, Finkelstein noted, after having presented the above measurement definition, "It is required that [mapping] *M* be a well-defined operational procedure" (p. 105).

Finkelstein not only shifted the focus from uniqueness problems back to representational problems; he also discussed problems of measurement that are very prominent in instrument and control engineering, but were not accounted for in the representational theory of measurement, namely, measurement errors and uncertainty: "The problems of errors in measurement scales have not been adequately discussed in the literature" (p. 110). He related this problem of uncertainty to the "perfection of the experimental apparatus and the number of the experiments carried out" (p. 110). By expressing the problem of error this way, he simply reflected the dominant view at that time, the traditional approach, which assumes that errors can be reduced by increasing precision (see section 2.1 in this chapter). While Finkelstein expected "a review of the problems . . . in the forthcoming continuation of the work of Krantz, Luce, Suppes and Tversky" (p. 110), it would actually take 20 years before such an account would be developed, the uncertainty approach, but it would not be by Krantz and coauthors.

Norman H. Anderson's criticism (1981, pp. 350–351) represents nicely the general critique of the axiomatic theory of measurement that exists outside mathematical psychology. His criticism consists of the description of the three limitations that characterize the axiomatic approach. The first limitation is that it leaves out the question of how the mathematical structures gain their empirical significance in actual practical measurement. Second, the axiomatic approach lacks concrete measurement procedures, devices, and methods. The representation theorem is nonconstructive: Although the theorem may imply that a scale exists, it does not provide a constructive way to know which one. And third, the axiomatic approach applies only to error-free data; it says nothing about handling the response variability in real data. An empirical representational theory of measurement should address these three issues.[19]

To arrive at an empirically significant representational theory, recall that the axiomatic approach arose as a reaction to Campbell's requirement for fundamental measurement that one should employ the procedure of additivity. For nonadditive properties, however, Campbell discussed the possibility of derived measurement, which requires an empirical law relating the relevant properties. Moreover, he noted that "the number of non-additive properties recognized in physics greatly exceeds that of additive, and there are many variants of the indirect [derived] process" (Campbell 1940, p. 341). While the practice of derived measurement is overrepresented, the axiomatic approach addresses only fundamental measurement.[20] For the exploration of the possibilities of an empirical theory of measurement, it is to be expected that one could better focus on derived measurement.

To do so, let us first have a closer look at what derived measurement is. An extensive discussion of derived measurement can be found in Brian Ellis's *Basic Concepts of Measurement* (1966), which is actually based on Campbell's *Physics* (1920) account of derived measurement.

Derived measurement is measurement by the determination of a constant in a "numerical law." Derived measurement is possible only if:

a) the systems which possess p (the quantity to be measured) possess other quantities $q, r, s, \ldots$, which are independently measurable;
b) there is a relation of $q, r, s, \ldots, f(q, r, s, \ldots)$ in which a constant c occurs which varies from system to system;
c) this system-dependent constant c is such that when the system is arranged in the order of c, they are also arranged in the order of p (see Ellis 1966, p. 56).

So the magnitude of property p is system-dependent and determined by the properties $q, r, s, \ldots$, that appear in the relation.

The occurrence of a constant in the numerical relation (the ascertainment of constancy is based on empirical investigation) reveals the (empirical) fact that there exists an invariant empirical relation, where this invariance is system-related. If it appears that the constant c is system-independent, so is "universal," one has found a universal law between $q, r, s, \ldots$.

This account of derived measurement is very similar to the account of measurement in metrology, where the earlier discussed fundamental measurement is hardly discussed at all. This metrological account can be found in one of the central documents in metrology, the *Guide to the Expression of Uncertainty in Measurement* (JCGM 100 2008, p. 8):

> In most cases, a measurand Y is not measured directly, but is determined from N other quantities $X_1, X_2, \ldots, X_N$ through a functional relationship $f : Y = f(X_1, X_2, \ldots, X_N)$,

where f is called the measurement function, and where $X_1, X_2, \ldots, X_N$ are called the input quantities and Y the output quantity.

To arrive at this function one first needs to model the measurement process, which means finding a "mathematical relation among all quantities *known* to be involved in a measurement" (JCGM 200 2012, p. 32; emphasis added). If data indicate that f does not model the measurement to the degree imposed by the required accuracy of the measurement result, additional input quantities must be included in f to reduce this inaccuracy (see JCGM 100 2008, p. 9).

The problem with this modeling strategy, however, is that "accuracy of measurement" is not a straightforward measure to validate a measurement model. The reason is that accuracy is defined with respect to the true value of the measurand: "closeness of agreement between a measured quantity value and a true quantity value of a measurand" (JCGM 200 2012, p. 21). But a true value is a value "that would be obtained by a perfect measurement," which is "only an idealized concept" (JCGM 100 2008, p. 50); therefore, "true values are by nature indeterminate" (p. 32).

> The first step in making a measurement is to specify the measurand—the quantity to be measured; the measurand cannot be specified by a value but only by a description of a quantity. However, in principle, a measurand cannot be *completely* described without an infinite amount of information. (JCGM 100 2008, p. 49)

To reflect this incomplete knowledge of the measurand, it is within current metrology generally acknowledged that measurement should be expressed in terms of uncertainty. But the usage of the concept of "uncertainty," instead of "error," is "relatively new in the history of measurement" (p. viii). It only started in 1977, when the "world's highest authority in metrology" (p. vi), the Comiteé International des Poids et Mesures, recognized "the lack of international consensus on the expression of uncertainty in measurement" (p. vi) and therefore requested the International Bureau of Weights and Measures to address the problem in conjunction with the national standards laboratories and to make a recommendation. From a questionnaire prepared by the Bureau and distributed to 32 national metrology laboratories, the consensus arose, by early 1979, that "it was important to arrive at an internationally accepted procedure for expressing measurement uncertainty and for combining individual uncertainty components into a single total uncertainty" (p. vi). But there was no consensus on the method to be used. Therefore the bureau installed a Working Group on the Statement of Uncertainties, which came with the following "Recommendation INC-1 (1980) Expression of experimental uncertainties":

> The uncertainty in the result of a measurement generally consists of several components which may be grouped into two categories according to the way in which their numerical value is estimated:

A. those which are evaluated by statistical methods,
B. those which are evaluated by other means. (JCGM 100 2008, p. ix)

Following this recommendation the *Guide to the Expression of Uncertainty in Measurement* (GUM) was developed, of which the first version was published in 1995. Due to the influence this guide had and still has in metrology, the new developed uncertainty approach became also to be known as the "GUM approach."

The uncertainty approach has consequences for the way measurement models are built; therefore modeling is specifically addressed in the *Guide*.[21] Models should be built "to express what is learned about the measurand" (JCGM 104 2009, p. 3). Uncertainty, defined as the "non-negative parameter characterizing the dispersion of the quantity values being attributed to a measurand, based on the information used" (JCGM 200 2012, p. 25), "reflects the lack of exact knowledge of the value of the measurand" (JCGM 100 2008, p. 5), and consists of several components, that is, "sources of uncertainty." The evaluation of these components should improve the "quality" of the measurement result.

The *Guide* lists the following components of uncertainty:

a) incomplete definition of the measurand;
b) imperfect realization of the definition of the measurand;
c) the sample measured may not represent the defined measurand;
d) inadequate knowledge of the effects of environmental conditions on the measurement or imperfect measurement of environmental conditions;
e) personal bias in reading analogue instruments;
f) finite instrument resolution or discrimination threshold;
g) inexact values or measurement standards and reference materials;
h) inexact values of constants and other parameters obtained from external sources and used in the data-reduction algorithm;
i) approximation and assumptions incorporated in the measurement method and procedure;
j) variations in repeated observations of the measurand under apparently identical conditions. (JCGM 100 2008, p. 6)

These sources are not necessarily independent, and some of sources (a) to (i) may contribute to source (j). Moreover, an unrecognized causal factor, although not being taken into account in the uncertainty evaluation of the measurement result, will nevertheless contribute to measurement error. It is also acknowledged that "blunders in recording or analyzing data can introduce a significant unknown error in the result of a measurement" (JCGM 100 2008, p. 8), but the *Guide*'s measures of uncertainty are "not intended to account for such mistakes" (p. 8).

According to Recommendation INC-1, these uncertainty components should be grouped into two categories based on their method of evaluation, "Type A" and "Type B." Type A evaluation is an "evaluation of a component of measurement uncertainty by a statistical analysis of measured quantity values obtained under defined measurement conditions" (JCGM 200 2012, p. 26). The defined measurement conditions under which the measurements are obtained for the statistical analysis in the Type A evaluation of measurement uncertainty range from repeatability to reproducibility conditions:

1. Repeatability conditions: same measurement procedures, same operators, same measuring system, same operating conditions and same location, and replicate measurements on the same or similar objects over a short period of time (see JCGM 200 2012, p. 23).[22]
2. Intermediate precision conditions: same measurement procedure, same location, and replicate measurements on the same or similar objects over an extended period of time, but may include other conditions involving changes (p. 24).
3. Reproducibility conditions: different locations, operators, measuring systems, and replicate measurements on the same or similar objects (p. 24).

These measurement conditions ranging from conditions of repeatability (1) to conditions of reproducibility (3) can be seen as different levels of control ranging from a strong level of laboratory conditions to a weaker level of field conditions.[23] But this shift from laboratory conditions to field conditions implies also a shift of emphasis from Type A evaluations to Type B evaluations.

Type B evaluation is an "evaluation of a component of measurement uncertainty determined by means other than a Type A evaluation of measurement uncertainty" (p. 26). These other means are based on information:

- associated with authoritative published quantity values,
- associated with the quantity value of a certified reference material,
- obtained from a calibration certificate,
- about drift,
- obtained from the accuracy class of a verified measuring instrument,
- obtained from limits deduced through personal experience. (JCGM 200 2012, p. 26)[24]

So these Type B evaluations are based on "other kinds" of information, such as authoritative and certified sources of information and professional judgment.

In the uncertainty approach the input quantities $X_1, X_2, \ldots, X_N$ are categorized into two groups, directly determined quantities and externally determined quantities:

- quantities whose values *and uncertainties* are directly determined in the current measurement. These values and uncertainties may be obtained from, for example, a single observation, repeated observations, or judgment based on experience, and may involve the determination of corrections to instrument readings and corrections for influence quantities, such as ambient temperature, barometric pressure, and humidity;
- quantities whose values *and uncertainties* are brought into the measurement from external sources, such as quantities associated with calibrated measurement standards, certified reference materials, and reference data obtained from handbooks. (JCGM 100 2008, p. 9)

The distinction between Type A and Type B components implies two different stages of modeling, a Type A stage and a Type B stage. A Type A stage exploits the measurement conditions under which the observations are obtained: "If all of the quantities on which the result of a measurement depends are varied, its uncertainty can be evaluated by statistical means" (JCGM 100 2008, p. 7). These conditions reflect the conditions of a laboratory, and so the conditions of experimental control. The degree of control is of course different for each category of measurement conditions, where the repeatability conditions represent most control and reproducibility conditions least control. A Type B stage depends on "skilled judgement" and external sources, which may be used as an additional pool of information about whether the model is complete. Combined, both stages lead to the following strategy of modeling:

> Because the mathematical model may be incomplete, all relevant quantities should be varied to the fullest practicable extent so that the evaluation of uncertainty can be based as much as possible on observed data. Whenever feasible, the use of empirical models of the measurement founded on long-term quantitative data, and the use of check standards and control charts that can indicate if a measurement is under statistical control, should be part of the effort to obtain reliable evaluations of uncertainty. The mathematical model should always be revised when the observed data, including the result of independent determinations of the same measurand, demonstrate that the model is incomplete. (JCGM 100 2008, p. 7)

When the model is considered to be complete, and the uncertainties of each of the input quantities are known, the uncertainty of y, the "estimate" of the measurand Y, can be given, and thus the result of the measurement is obtained by appropriately combining the uncertainties of the input estimates $x_1, x_2, \ldots, x_N$. This combined uncertainty of estimate y, denoted by $u_c(y)$, is the positive square root of the combined variance $u_c^2(y)$ that is given by

$$u_c^2(y) = \sum_{i=1}^{N} \left(\frac{\partial f}{\partial x_i}\right)^2 u^2(x_i),$$

where f is the measurement function (see JCGM 100 2008, p. 19). The partial derivatives $\partial f/\partial x_i$ are called "sensitivity coefficients" and describe how the output estimates y varies with changes in the values of the input estimates of $x_1, x_2, \ldots, x_N$. In particular, the change in y produced by a small change Δx_i in input estimate x_i is given by $(\Delta y)_i = (\partial f/\partial x_i)(\Delta x_i)$.

Knowledge of these sensitivity coefficients is thus essential in measurement. If the function f is known, they can be "calculated" from that function. If not, they have to be determined in an empirical way: "sensitivity coefficients are sometimes determined experimentally: one measures the change in Y produced by a change in a particular X_i while holding the remaining input quantities constant" (p. 20). The *Guide* gives no guidance when these experiments are not possible, that is, when one cannot impose ceteris paribus conditions. Therefore one could conclude that the "GUM approach" is an account of laboratory measurement.

2.4. Measurement outside the Laboratory

To arrive at an account for measurement outside the laboratory, one could go to one of the field sciences, such as econometrics, and see what kind of measurement methodology it has developed. In particular the problem of arriving at knowledge about the sensitivity coefficient $\partial f/\partial x_i$ was in an advanced form dealt with in the works of the Norwegian econometrician Trygve Haavelmo. Haavelmo's work therefore will be dealt with in chapter 4, "The Problem of Passive Observation."

While Type A evaluations outside the laboratory can use observations obtained under "reproducibility conditions," the sources of information for Type B evaluations are much more problematic for the social field sciences. In metrology these sources of information include "authoritative published quantity values," "certified reference materials," and "calibration certificates."

In social sciences, such as in economics, these authoritative sources of information only exist for Type A quantities, like those from the US Bureau of Economic Analysis. There are no institutions in social science with the same level of authority, like the International Bureau of Weights and Measures (BIPM) in metrology, for publishing Type B quantity values. Moreover, there are no certifications for reference materials and calibrations in social science, because this would presume the existence of accepted authorities for providing these certificates.

There are, however, proposals in economics to allow "stylized facts" to be these authoritative Type B quantity values. For example, Thomas Cooley and Edward Prescott (1995, p. 3) propose to consider the stylized facts of economic growth, which are the "striking empirical regularities both over time and across countries", as "benchmarks of the theory of economic growth," so they can be used to calibrate business cycle models. The stylized growth facts they propose are the following:

> 1. Real output grows at a more or less constant rate.
> 2. The stock of real capital grows at a more or less constant rate greater than the rate of growth of the labor input.
> 3. The growth rates of real output and the stock of capital tend to be about the same.
> 4. The rate of profit on capital has a horizontal trend. (Cooley and Prescott 1995, p. 3)

But these facts lack authority. To take them as empirical facts is controversial. As Robert Solow (1970, p. 2) remarked, "There is no doubt that they are stylized, though it is possible to question whether they are facts."

As a result, the only available sources of information left over for Type B evaluations are "information about drift," "the accuracy class of a verified measuring instrument," and "personal experience." Drift is an undesirable gradual deviation of an instrument output over a period of time that is unrelated to changes in input operating conditions or load. An instrument is said to have no drift if it reproduces the same readings at different times for same variation in measured quantity. Drift is caused by wear and tear, high stress developed at some parts, and so on. Drift relates to material instruments, and so does not provide relevant information about the validation of a model.

If one reads "model" for "measuring instrument" (as proposed in Boumans 2005), then the second source of information, the accuracy class of a verified measuring instrument, can be considered as the accuracy class of an empirically tested model. But this step is regressive, and leads us back to the original problem of evaluating the measurement model: the accuracy class of the measurement model is determined by the two types of evaluations, Type A and Type B, for which the same kinds of information sources are applicable or not. I discuss this in more detail subsequently.

It thus seems that the only available source of information for Type B evaluation in social field science is "professional skill":

> The proper use of the pool of available information for a Type B evaluation . . . calls for insight based on experience and general knowledge, and is a skill that can be learned with practice. (JCGM 100 2008, p. 12)

Although a Type B evaluation is meant to be a "scientific judgment" (JCGM 104 2009, p. 5), it is apparently more subjective than a Type A evaluation.

However, as will be shown here, it is possible to develop a more objective Type B evaluation for measurement in social field science. To arrive at such an evaluation, one has to use a model-based evaluation. This model-based evaluation is actually a specification of the accuracy class of a measurement model based on the determination of the validity of the measurement model according to a particular Type B strategy.

The basic idea of an objective Type B evaluation can be briefly summarized as follows: When modeling the measurement process, one should include every potential input quantity, X_i, suggested by "theory," "experience," and "general knowledge," irrespective of whether we have (enough) observations to assume its potential influence. Subsequently the validity of this encompassing model should be tested—which makes this approach an objective evaluation. The model may still be incomplete, but the tests will tell whether a significant input quantity is still missing or whether the input quantities not included in the model are negligible. To deal with input quantities that are not measurable or for which we have not enough observations for a Type A evaluation, the proposal is to use a "gray box" modeling approach instead of a "white box" modeling approach.

A white-box model is a set of causal-descriptive statements on how a real system actually operates in some aspect. Testing this kind of model involves taking each relationship individually and comparing it with observations of the real system. Type A evaluations assume such a white-box approach, because we have the observations (statistical data) available to evaluate each relationship individually. As will be shown in what follows, a Type B evaluation does not require this kind of model. For Type B evaluations the model can be a gray-box model. A gray-box model is a modular designed model, where the modules can be considered as black boxes. Testing this kind of model does not require having observations for each individual relationship.

To clarify this distinction between white-box, gray-box, and black-box models and the different kinds of testing they require, Yaman Barlas's (1996) distinction between three stages of model validation is useful. Barlas's motivation for making this distinction is that system dynamics has "often been criticized for relying too much on informal, subjective and qualitative validation procedures" (p. 183). His purpose is to discuss more objective procedures for model validation, without implying that model validation "can be cast as *entirely* formal/objective constructs and procedures" (p. 183). Essential for his discussion is that validation of a model cannot be divorced from its purpose, which implies that validation and model construction are closely related.

The three stages of validation Barlas distinguishes are (1) "direct structure tests," (2) "structure-oriented behaviour tests," and (3) "behaviour pattern tests." *Direct structure tests* assess the validity of the model structure, by direct comparison with knowledge about the real system structure. This involves taking each relationship individually and comparing it with available knowledge about the real system. This knowledge can be theoretical as well as empirical. Examples of direct structure tests are structure and parameter confirmation tests, direct extreme-condition tests, and dimensional consistency tests.

Structure-oriented behavior tests assess the validity of the structure indirectly, by applying certain behavior tests on model-generated behavior patterns. These tests involve simulation, and can be applied to the entire model, as well as to isolated submodels of it. "These are 'strong' behavior tests that can help the modeler uncover potential structural flaws" (p. 191). Examples are extreme-condition tests, behavior sensitivity tests, modified-behavior predictions, boundary adequacy tests, phase-relationship tests, qualitative features analyses, and Turing tests. An extreme-condition test is a stress test; it involves assigning extreme values to selected parameters and comparing the model-generated behavior of the observed behavior of the real system under the same extreme condition. A behavior sensitivity test consists of determining those parameters to which the model is highly sensitive, and observing whether the real system is exhibiting similar sensitivity to the corresponding parameters. Modified-behavior prediction is comparing the behavior of a simulated model with structural modifications with behavior of a modified version of a real system. A phase-relationship test uses the phase relationship between pairs of variables in the model, obtained as a result of a simulation. If certain of these phase relationships contradict the phase relationships that are observed from the real system, this may indicate a structural flaw of the model. A qualitative feature test is what in econometrics is called a characteristics test (Kim, De Marchi, and Morgan 1995; Boumans 2005, pp. 89–92) and compares the qualitative features of the simulated behavior with the observed behavior. In a Turing test, experts are presented with a shuffled collection of real and simulated output behavior patterns, and asked if they can distinguish between these two types of patterns. Interestingly Barlas (1996) suggested that if experts do detect significant differences, then "they are interviewed with the purpose of uncovering the structural flaws in the model that may have caused the differences" (p. 192). This "elicitation" of experts will be discussed in more detail in chapter 6.

Behavior pattern tests do not evaluate the validity of the model structure, either directly or indirectly, but measure how accurately the model can reproduce the major behavior patterns exhibited by the real system. They are tests based on pattern prediction, not limited to point prediction. Patterns include periods, frequencies, trends, phase, lags, and amplitudes.

For white-box models all three stages are equally important, while for black-box models only the last stage matters. Barlas (1996) does not refer to gray-box models. Although Barlas emphasizes that structure-oriented behavior tests are designed to evaluate the validity of the model structure, his usage of the notion of structure with respect to these kinds of tests allows for a notion of structure that is not limited to realistic descriptions of real systems; it also includes other kinds of arrangements, like modular organizations. Structure-oriented behavior tests are also adequate for the validation of modular-designed models, and for these models the term "structure" refers to the way the modules are assembled.

A module is a self-contained component (to be treated as a black box) with a standard interface to other components within a system (White 1999, p. 475). These modular-designed models—in line with the labeling of the other two types of models—will be called "gray-box models" and should pass the structure-oriented behavior tests and the behavior pattern tests.

This concept of a gray-box model and the way it should be validated is useful for developing a more objective Type B uncertainty evaluation for measurement in social field science. The first step is to acknowledge that the measurement model does not have to be a white-box model; that is, the model does not need to be a homomorphic mapping of the measurand. White-box models are required to answer "why" questions, but for "how much" questions we can use gray-box models (see Boumans 2006).

The result of this step is that a measurement model does not need to be validated with direct structural tests, because structure-oriented behavior tests and behavior pattern tests suffice. These validation tests are based on evaluations of whether simulated model output patterns match real data patterns within some specified range of accuracy, without questioning the validity of the individual relationships that exist in the model. These patterns are the result of the *interplay* of the relations between the modules and the specific characteristics of the modules. Structure-oriented behavior tests therefore provide feedback on these interplays. The individual modules are validated by behavior pattern tests. In other words, a measurement model is a specific assemblage of black-box models, each of which is validated by behavior pattern tests, and of which the assemblage—"structure"—is validated by structure-oriented behavior tests.

The consequence of this model design is that the measurement model does not need to be a complete causally descriptive representation and that the modeling does not require detailed statistical knowledge about the complete set of the input quantities. Notwithstanding these weaker requirements on knowledge of the system and available observations, strong validation tests exist that are able to identify and even to estimate the magnitude of the uncertainty of neglected, ignored, or unknown influence quantities. The application of this kind of modeling and validation enables more objective Type B evaluation strategies.

2.5. Conclusions

Actually there exists no "theory" to account for measurement practices outside the laboratory. The heuristic rule that has been applied in this chapter is the Paul principle: "Question everything; keep what is good." The result is more a methodology than a theory of measurement outside the laboratory. In other words, we arrived at a methodology that prescribes a specific way of building models (modular design) and a specific way of validating models (structure-oriented behavior tests).

The only existing systematic account of measurement is the axiomatic theory of measurement, but, as we have seen, it does not account for measurement of field phenomena. This chapter has taken the representational theory of measurement as a starting point, extending it in such a way as to arrive at a methodology adequate for measuring social phenomena outside the laboratory. Therefore we traveled presumptuously across several disciplinary boundaries to study their specific practices of measurement and to keep what is useful for our own aims.

What is striking in these practices is that there are no explicit accounts for measurement, but that the main methodological principles are interwoven in "guides," "vocabularies," and scattered methodological articles. This not only applies for the subject of this chapter, but will also be the case for the subjects of the following chapters. In studying "science outside the laboratory," the "naturalistic turn" (that is, studying practices instead of theories) is a natural way to proceed.

Measurement is the assignment of numerals to a property of objects or events—the "measurand"—according to a rule with the aim of generating reliable information about these objects or events. The central measurement problem is the design of rules so that the information is as reliable as possible. To arrive at reliable numbers for events or objects, the rules have to meet specific requirements. The nature of these requirements depends on the nature of the event or object to be measured and on the circumstances in which the measurements will be made.

According the representational theory of measurement, the most appropriate rule for the measurement of social phenomena is a homomorphism that maps the relations between the relevant features of the measurand into a model. This model is a representation of the empirical relational structure. The implicit consequence of the homomorphism requirement is that for the measurement to be reliable, the model needs to be as complete as possible. Completeness means here that the model encompasses all possible influences that may affect the measurand. But outside the laboratory our knowledge about those influences is very much dependent on how far they have been observed, that is, appear in the statistics. These observations can be evaluated in an objective way by the application of the relevant statistical methods and techniques—the Type A evaluation.

But statistics is not sufficient to make a model complete. Additional knowledge is required, which is often provided by an expert with skilled and experienced knowledge about the measurand and the methods and techniques to measure it.

This additional expert knowledge makes the validation of the representation less objective. In system dynamics, however, several validation tests are used that, combined with specific model designs, could be used to make the assessment of incomplete models more objective. This chapter suggeststhat a gray-box model validated by structure-oriented behavior tests, with its constituent modules validated by behavior pattern tests, is a sufficiently accurate representation to make the measurements sufficiently reliable.

The consequence of this is that the measurement model does not have to be a homomorphism of the structural relations describing the measurand.

Notes

1. This view, expressed in his 1879 article "Psychometric Experiments," became the motto of the Francis Galton Laboratory of National Eugenics.
2. Dutch physicist who was awarded the Nobel Prize for Physics in 1913 for his investigations on the properties of matter at low temperatures, which led, inter alia, to the production of liquid helium. The laboratory in which these low-temperature experiments were conducted had "door meten tot weten" as its motto. It was the title of his inaugural lecture on the occasion of his appointment as professor at the University of Leiden in 1882 (Laesecke 2002).
3. This statement was the motto of the Cowles Commission. The Cowles Commission for Research in Economics (which became in 1952 the Cowles Foundation) was founded in 1932 by Alfred Cowles and had as its purpose "the conduct and encouragement of research in economics, finance, commerce, industry, and technology, including problems of the organization of these activities, and of society in general. Its approach is to encourage and extend the use of logical, mathematical, and statistical methods of analysis" (Cowles Commission 1952, p. v). Measurement was used in its broadest sense of systematic observation. "It ranged from mere classification of objects, through the establishment of preferences between objects, to the construction of numerical scales and measures" (Cowles Commission 1952, p. 1). Nancy Cartwright opens her *Nature's Capacities and Their Measurement* (1989) with the same motto. She has studied the econometrics of the Cowles Commission for the same reason as I have, particularly in my chapter 4: "not because of either the successes or the failures of econometrics as a science; rather because of the philosophic job it can do" (Cartwright 1989, p. 7).
4. See Boumans and Davis 2010, chapter 1, for a more detailed discussion of what this received view entails.
5. Metrology is the shared view on measurement of eight international metrological organizations: the International Bureau of Weights and Measures (BIPM), the International Electrotechnical Commission (IEC), the International Federation of Clinical Chemistry and Laboratory Medicine (IFCC), the International Laboratory Accreditation Cooperation (ILAC), the International Organization for Standardization (ISO), the International Union of Pure and Applied Chemistry (IUPAC), the International Union of Pure and Applied Physics (IUPAP), and the International Organization of Legal Metrology (OIML). Their shared view can be found in the publications of the Joint Committee for Guides in Metrology (JCGM). One of the publications they prepare and approve is the *International Vocabulary of Basic and General Terms in Metrology* (JCGM 200 2012), in which metrology is defined as "science of measurement and its applications" (p. 16).
6. In metrology, the uncertainty approach is developed in the subsequent editions of the *Guide*. Beside the *Vocabulary*, the editions of the *Guide* are the key publications of the Joint Committee for Guides in Metrology.
7. Boumans (2004) discusses how the concept of a model and the practice of modeling in economics originate in Maxwell's method of analogy.
8. The other members in 1940 were C. S. Myers (vice-chairman), R. J. Bartlett (secretary), H. Banister, F. C. Bartlett, W. Brown, K. J. W. Craik, J. Drever, J. Guild, R. A. Houstoun, J. O. Irwin, G. W. C. Kaye, S. J. F. Philpott, L. F. Richardson, J. H. Shaxby, T. Smith, R. H. Thouless, and W. S. Tucker. See Michell 1999, pp. 143–155, for a detailed history of this committee.
9. Campbell made a distinction between "numbers" and "numerals": Numerals are representations of numbers: "anything which thus represents a number is a 'numeral'" (Campbell 1920, p. 269). This usage became standard.

10. Density became the standard example for derived measurement, like length for fundamental measurement.
11. Contributors were Peter Caws, C. West Churchman, Coombs, Donald Davidson, E. J. Gumbel, Paul Kircher, Luce, Henri Margenau, Jacob Marschak, John L. McKnight, Karl Menger, Arthur Pap, Stevens, Suppes. They were "chosen because it was known that they had different viewpoints on the meaning and significance of measurement" (Churchman and Ratoosh 1956, p. v).
12. Stevens (1939) discusses seven principles, but I have selected here only those principles that are relevant for the discussion of measurement.
13. They also preferred using the term "model" instead of "scale." See Chapter 1 for their view according to which an inference from a mathematical model can have the same role as an experiment.
14. This period of the history of measurement is closely connected to the development of game theory. For example, Raiffa and Luce wrote a book *Games and Decisions* (1957), which was "the first non-mathematical exposition that made the theory of games accessible to the broad community of social scientists" (Kuhn 2004, ix). See Heukelom 2010 and Erickson 2013 for more historical details. It should be emphasized, however, that the axiomatic approach itself entered measurement theory well before the 1950s. In 1901, Otto Ludwig Hölder published a paper on "the axioms of quantity and the theory of measurement" (translated by Michell and Ernst 1996, 1997); see for its history Michell 1993. Probably the first example of the axiomatic approach in economics is Frisch (1926), in which three axioms define utility as a quantity (see Boumans 2012). See Moscati 2013 for a history of the measurement of utility.
15. The meaning of "full" in this definition is that the domain of the numbers is the set of all real numbers.
16. The symbol $\circ$ denotes the composition of functions.
17. The theory of extensive measurement is a set of assumptions formulated in terms of an ordering and a "concatenation operation."
18. Translated by M. J. Scott-Taggart and Brian Ellis from Ernst Mach, *Die Principien der Wärmelehre* (Leipzig, 1896), pp. 39–57, and published as an appendix in Ellis 1966.
19. For a more recent critique from metrology, see Frigerio, Giordani, and Mari (2010, p. 126): "RTM is *too abstract* to be useful in a scientific context where since a long time measurement instrumentation has been designed, built and properly used."
20. Derived measurement is scarcely discussed in the *Foundations of Measurement*. Only the subject index of volume 1 mentions "derived measures," and in that volume derived measurement is only discussed with respect to the different meanings it has among Krantz et al., Campbell (1920), a book on dimensional analysis by J. Palacios, and Ellis (1966).
21. As part in a series of JCGM documents, "Evaluation of measurement data," there will appear a special document on modeling, but it is still in preparation and so not yet published (verified on 26 June 2014).
22. Note that this condition is similar to Stevens's second "operational principle."
23. Glenn Harrison and John List (2004, pp. 1013–1014) discuss a taxonomy for experiments in economics, which is not based on different levels of control but on different factors that determine the field context of an experiment:
    - A *conventional lab experiment* is one that employs a standard subject pool of students, an abstract framing, and an imposed set of rules.
    - An *artifactual field experiment* is the same as a conventional lab experiment but with a nonstandard subject pool.
    - A *framed field experiment* is the same as an artifactual field experiment but with field context in either the commodity, task, or information set that the subjects can use.
    - A *natural field experiment* is the same as a framed field experiment, but the environment is one where the subjects naturally undertake these tasks and where the subjects do not know that they are in an experiment.

24. This is the most recent description of the pool of information for Type B evaluations. A few years earlier the GUM (JCGM 100 2008, p. 11) listed the following, similar items:
    - Previous measurement data
    - Experience with or general knowledge of the behavior and properties of relevant materials and instruments
    - Manufacturer's specifications
    - Data provided in calibration and other certificates
    - Uncertainties assigned to reference data taken from handbooks

3

Calculus of Observations

Everybody believes in the exponential law of errors: the experimenters, because they think it can be proved by mathematics; and the mathematicians, because they believe it has been established by observation.

—Henri Poincaré[1]

3.1. Introduction

There are many quantitative field sciences where a statistical model cannot provide a complete representation of the object of study and where one needs additional sources of knowledge based on expert judgments. In this chapter this broader body of knowledge that includes both statistical models and expert judgments is called a "calculus of observations." This chapter discusses and compares different "calculi of observations," used in various field sciences, that aim at inferring from a sample of observations knowledge of facts about phenomena, particularly when these phenomena are observed outside the laboratory. To facilitate this comparison of calculi coming from these so different kinds of field sciences, the problem field will be narrowed down to one specific problem, namely that of accuracy.

The problem of accuracy can be explained (and simplified) as follows. Assume that one wishes to measure a variable x, and the value of x has to be inferred from a set of available observations $y_i (i = 1, \ldots, n)$, which inevitably involve noise ε_i. One can then represent this relation between these observations, y_i, the measurand, x, and noise, ε_i, by the following equation:[2]

$$y_i = f(x) + \varepsilon_i. \tag{3.1}$$

This equation will be referred to as the "observation equation." Accuracy will be attained by the reduction of noise. In a laboratory, where we can control the environment, this can be done by "cleaning" the environment, but outside the laboratory, accuracy has to be obtained by a "calculus of observations."

I borrow the term "calculus of observations" from the title of a textbook written by the English mathematician Edmund T. Whittaker and the Canadian mathematician George Robinson, and first published in 1924 (Whittaker and Robinson

1924, 1944). The book contains a course of lectures Whittaker had given during the years 1913–1923 to students at the Mathematical Laboratory of the University of Edinburgh, which he had founded in 1913. The 1924 textbook's chapter titles indicate the kind of topics that were studied in the laboratory: interpolation, central-difference formulas, determinants and linear equations, numerical solution of algebraic and transcendental equations, numerical integration, normal frequency distribution, method of least squares, practical Fourier analysis, graduation, correlation, search for periodicities, and numerical solutions of differential equations.

Although Whittaker's mathematical laboratory had the appearance of a physical laboratory, most central to the work of the mathematical laboratory was the development of numerical methods.[3] Whittaker had developed a collection of new techniques for solving problems numerically and had worked out efficient routines for undertaking this numerical work: the procedure to be followed during each calculation was governed by the use of a special printed form. For example, if a collection of data points was to be given a Fourier analysis, the period was divided into 12 standard intervals of 30° each and then analyzed using a standard form containing boxes for each stage of the calculation. The form ensured that the computer (human, not machine) recorded each of the steps of his or her calculation and undertook a self-check (also recorded) to ensure that he or she had not made any major computational errors.

Sheets of computing papers were also designed for the graduation method. A part of such a computing sheet was shown in the 1944 book (see figure 3.1). The products in each column were "computed either by arithmometer or, in three place work, by Crelle's Tables" (1944, p. 316). According to the graduation formula the diagonal entries should be summed up; bold numbers in the figure show an example. To avoid errors of adding the wrong numbers, the use of a special stencil was recommended: "distraction of the eye may be avoided by the sum of a V-shaped stencil, which leaves exposed at any time only the particular converging diagonals required" (1944, p. 316). Checks during the computations were also provided.

Andrew Warwick (1995), who discusses Whittaker's laboratory, argues that mathematical calculation and precision measurement are similar activities: "the activity of calculating is neither more certain in outcome, nor more independent of the physical world, than is the activity in the laboratory" (p. 313). A consequence of this similarity is that reliability is gained in a similar way. When a novel phenomenon is first measured, to make people agree on the result, standardized procedures and technologies are developed to make the new result credible, such as making it reproducible. Because the answer to a computational problem is not given in advance, to make people agree on the outcome of a calculation, methods have to be devised, in a similar way as in a laboratory, for ensuring that agreement can be reached. "By developing a disciplined routine for calculating using the material technology of pencil and paper we have improved the reliability of the practice"

1.	1.	−1.	−3.	−6.	−5.	−1.	10.	27.	39.	31.	−17.	−103.	−211.	−236.	−46.	493.	1338.	2237.	2771.	x.	u_x	Δ^3	u'_x	Δ^3
																					...			
1	1	−1	−2	−4	−4	−1	7	19	28	22	−12										718			
1	1	−1	−2	−3	−3	−1	6	17	25	20	−11	−70									649			
1	1	−1	−2	−3	−3	−1	6	16	23	18	−10	−64	−124								589			
1	1	−1	−2	−3	−3	−1	5	15	21	17	−9	−58	−114	−127							538			
0	0	0	−1	−2	−2	0	5	13	19	15	−8	−54	−105	−117	−23						496			
0	0	0	−1	−2	−2	0	5	13	18	14	−8	−50	−98	−109	−21	225					463			
0	0	0	−1	−2	−2	0	4	12	17	14	−7	−47	−93	−104	−20	213	595				439			
0	0	0	−1	−2	−2	0	4	11	17	13	−7	−46	−89	−100	−20	206	575	974			424			
0	0	0	−1	−2	−2	0	4	12	17	13	−7	−47	−91	−102	−20	209	584	990	1194	20	431	−69	419	−4
0	0	0	−1	−2	−2	0	4	11	16	13	−7	−44	−86	−97	−19	199	555	939	1133	1	409	142	422	−6
0	0	0	−1	−2	−2	0	4	12	17	13	−7	−46	−91	−101	−20	208	582	985	1189	2	429	−248	435	−5
0	0	0	−1	−2	−2	0	4	11	16	13	−7	−46	−89	−100	−19	205	572	969	1160	3	422	112	454	0
1	1	−1	−2	−3	−3	−1	5	14	21	16	−9	−57	−112	−125	−24	258	719	1217	1460	4	530	107	473	8
1	1	−1	−2	−3	−3	−1	5	14	20	15	−9	−55	−107	−119	−23	245	685	1160	1300	25	505	−99	487	3
0	0	0	−1	−2	−2	0	5	12	18	14	−8	−50	−97	−108	−21	223	622	1054	1272	6	459	23	496	0
0	0	0	−1	−2	−2	0	5	13	19	14	−8	−54	−105	−118	−23	243	677	1146	1383	7	499	−23	508	−2
1	1	−1	−2	−3	−3	−1	5	14	21	16	−9	−57	−111	−124	−24	256	713	1208	1458	8	526	−3	526	0
1	1	−1	−2	−3	−3	−1	6	15	22	17	−10	−61	−119	−133	−26	274	763	1293	1560	9	583	60	550	0
1	1	−1	−2	−3	−3	−1	6	16	23	18	−10	−63	−124	−139	−27	285	796	1348	1627	30	587	−74	578	−1
1	1	−1	−2	−3	−3	−1	6	16	23	18	−10	−64	−126	−140	−27	289	808	1367	1649	1	595	85	610	−3
1	1	−1	−2	−3	−3	−1	6	17	25	20	−11	−70	−137	−153	−30	314	877	1486	1793	2	647	−118	646	−5
1	1	−1	−2	−3	−3	−1	7	18	28	21	−11	−72	−141	−168	−31	325	907	1637	1854	3	669	67	685	0
1	1	−1	−2	−4	−4	−1	7	20	29	23	−13	−81	−167	−176	−34	363	1012	1714	2067	4	746	28	724	1
1	1	−1	−2	−4	−4	−1	8	21	30	24	−13	−82	−160	−179	−35	369	1031	1746	2106	35	760	−64	758	4
1	1	−1	−2	−4	−4	−1	8	21	30	24	−13	−84	−164	−184	−36	378	1055	1787	2156	6	778	4	787	6
1	1	−1	−2	−4	−4	−1	8	22	32	26	−14	−89	−175	−195	−38	402	1123	1902	2294	7	828	118	812	5
1	1	−1	−3	−4	−4	−1	8	23	33	26	−14	−91	−179	−200	−39	411	1147	1943	2344	8	846	−130	837	−2
1	1	−1	−3	−4	−4	−1	8	23	33	26	−14	−90	−176	−197	−38	406	1134	1920	2317	9	836	58	868	−5
1	1	−1	−3	−5	−4	−1	9	25	36	28	−16	−99	−193	−216	−42	445	1242	2104	2538	40	916	−14	910	−6
1	1	−1	−3	−5	−5	−1	10	26	37	30	−16	−103	−202	−226	−44	465	1296	2196	2649	1	956	−8	961	−8
1	1	−1	−3	−5	−5	−1	10	27	40	31	−17	−110	−214	−230	−47	493	1375	2329	2810	2	1014	−64	1016	−4
1	1	−1	−3	−5	−5	−1	11	29	42	33	−18	−116	−227	−254	−49	523	1450	2472	2982	3	1076		1069	
1	1	−1	−3	−6	−6	−1	11	31	44	35	−19	−122	−239	−268	−52	551	1538	2605	3142	4	1134		1112	
1	1	−1	−3	−6	−6	−1	11	30	44	35	−19	−121	−237	−265	−52	546	1524	2582	3115	45	1124		1141	
1	1	−1	−3	−6	−6	−1	12	31	45	36	−20	−125	−244	−273	−53	562	1568	2655			1156			
1	1	−1	−3	−6	−6	−1	12	31	45	36	−20	−125	−244	−273	−53	562	1668				1160			
1	1	−1	−3	−6	−6	−1	11	31	45	35	−19	−123	−241	−270	−53	555					1142			
1	1	−1	−3	−6	−6	−1	11	30	43	35	−19	−120	−235	−263	−51						1114			
1	1	−1	−3	−5	−5	−1	11	29	42	33	−18	−116	−226	−253							1072			
1	1	−1	−3	−5	−5	−1	10	27	40	32	−17	−110	−214								1016			
1	1	−1	−3	−5	−5	−1	9	26	37	29	−16	−102									946			
1	1	−1	−3	−4	−4	−1	9	23	34	27	−15										861			
																					...			

Figure 3.1 Computing Sheet for Graduation. Source: Whittaker and Robinson 1944, facing page 316.

(p. 316). Warwick shows that the reliability and utility of complicated calculations, just like the reliability and utility of precision measurements, can be improved and extended by developing new methods and technologies of computation.

A calculus of observations is a numerical combination of the values of the observations, $M[y_i]$, such that the estimate, $\hat{x}$, is as accurate as possible. So, the measurement value $\hat{x} = M[y_i]$ of a measurand x is accurate when $\hat{x}$ is very close to x.[4] One part of (but not equal to) accuracy is the closeness of the replicated measurements of the same measurand, which is called precision.[5] The example of Whittaker's mathematical laboratory shows that precision is acquired by specific procedures. Both numerical combination and the procedures that have to be followed when making the calculations are similar to the methods of control and intervention in a physical laboratory to reach accuracy and precision.

These procedures of the mathematical laboratory, however, are not only similar to the methods of control and intervention of the physical laboratory; they are also designed to replace them.[6] Control, as it were, is transposed from methods of intervention in the physical world to standardized procedures of computation. Outside the laboratory, where one cannot control the environment, accuracy has to be obtained by modeling in a specific way (cf. figure 1.2 in chapter 1).

Hence, to measure x, a model, denoted by M, has to be specified, for which the observations y_i function as input and $\hat{x}$, the estimation of x, functions as output:

$$\hat{x} = M\left[\, y_i; \alpha \right], \tag{3.2}$$

where α denotes the set of parameters a_i of the model. This equation is called the "measurement equation." The term "model" is used here in a very general sense; it includes all kinds of "calculi": econometric models, filters, graduations, and index numbers.

Substitution of the observation equation (eq. 3.1) into the measurement equation (eq. 3.2) shows what should be modeled (assuming that M is a linear operator, which is usually the case):

$$\hat{x} = M\left[f(x) + \varepsilon_i; \alpha\right] = M_x\left[x; \alpha\right] + M_\varepsilon\left[\varepsilon_i; \alpha\right].$$

From this algebraic exercise one can see that a necessary condition for $\hat{x}$ to be a measurement of x is that model M must be a representation of the observation equation, in the sense that it must specify how the observations are related to the measurand. Therefore we need both a representation of the measurand, M_x, and a specification of the observational errors, or noise, M_ε.

But representations of both measurand and the observational errors are only one part of an answer to the problem of the lack of control for obtaining accuracy in case of "passive observations" (see chapter 4). To see this, we first split the measurement error $\hat{\varepsilon}$ in two parts:

$$\hat{\varepsilon} = \hat{x} - x = M_\varepsilon + (M_x - x).$$

To explore how this measurement error is dealt with in the first instance, it is helpful to compare this measurement error with the "mean-squared error" of an estimator as defined in statistics:

$$E[\hat{\varepsilon}^2] = E[(\hat{x} - x)^2] = Var\hat{\varepsilon} + (E\hat{x} - x)^2.$$

The first term of the right-hand side of this expression of the mean-squared error, the variance of the measurement error, is a measure of "precision" (see next section) and the second term is called the "bias" of the estimator. Comparing the mean-squared error with the measurement error, one can see that the reduction—as much as possible—of the error term M_ε is aimed at by attempts to obtain precision.

To see what is needed to reduce the second error term, $M_x - x$, we have to split this term further by adding a new term that is relevant to assess accuracy, a norm, a "normal," that is, a standard value, N:

$$\hat{\varepsilon} = \mathrm{M}_\varepsilon + (M_x - N) + (N - x).$$

The reduction of the middle term, $M_x - N$, is in metrology called calibration, which is the establishment of the relationship between values indicated by a measuring system and the corresponding values realized by standards (see JCGM 200 2012, p. 28). For both precision and calibration, mechanical procedures have been developed that contribute to the "mechanical objectivity" of measurement.

Mechanical objectivity is used here in the same sense as defined by Theodore M. Porter (1995, p. 4): "It implies personal restraint. It means following the rules. Rules are a check on subjectivity: they should make it impossible for personal biases or preferences to affect the outcome of the investigation." Daston and Galison (2007, p. 121) describe mechanical objectivity aptly as "the insistent drive to repress the willful intervention of the [measurer], and to put in its stead a set of procedures that would, as it were, move nature to the page through a strict protocol, if not automatically." Mechanical objectivity thus reduces personal judgment as much as possible to emphasize procedures that have to be followed strictly.

The reduction of the third term $(N - x)$, standardization, as we will see below, cannot be done in a mechanically objective way. It involves prior theoretical assumptions of the measurand. But any theory is principally incomplete with respect to dealing with errors, in two different ways: a theory is both *unfinished* and *inexact* (see also chapter 4 for a more general discussion of inexactness of theories).

A theory is principally *unfinished* with respect to measurement errors because of the following reason: A standard is a numerical representation (a "numeral," as referred to by Campbell (1920); see chapter 2), whose values are obtained by measurement. But for measurement we need standards. Any empirical approach (Hoover 1994 characterizes this as "weak apriorism"; see also chapter 4) seems to show a kind of circularity: science is founded on measurement, which is founded on (theoretical) conceptualization of the measurand, which is founded on science. But in fact this process is not a closed circle, but is better captured by the term "epistemic iteration." Hasok Chang (2004), who coined this term, describes epistemic iteration as follows: In epistemic iteration we start by adopting an existing system of knowledge, with some respect for it but without any firm assurance that it is correct; on the basis of that initially affirmed system we launch inquiries that result in the refinement and even correction of the original system. Chang shows that it is this self-correcting process which justifies (retrospectively) successful courses of development in science, but does not provide any assurance of truth by reference to some indubitable foundation.[7]

James Bogen and James Woodward (1988) showed that any theory is also incomplete with respect to errors in the sense that any theory is *inexact*. According to them, well-developed scientific theories explain facts about phenomena but not facts about the observations that are the raw material of evidence for these theories. Many different factors play a role in the production of these observations, and the characteristics of such factors are heavily dependent on the peculiarities of, for example, the particular experimental design, detection devices, or data-gathering procedures that the investigator applies. Observations are idiosyncratic to particular experimental contexts. Moreover, the elements involved in the production of these observations will often be so disparate and numerous, and the details of their interactions so complex, that it is impossible to construct a theory that would allow us explain their occurrence or trace in detail how they combine to produce specific bits of data. Phenomena, by contrast, are not idiosyncratic to specific experimental contexts. Bogen and Woodward argue that phenomena have stable, repeatable characteristics that can be detectable by means of a variety of different procedures; however, these procedures may yield quite different data in different contexts. Addressing errors will differ from case to case and depend upon the effects of many different conditions peculiar to the subject under investigation, such as the experimental design and the equipment used. "The factors contributing to observational error, and the means for correcting for it in different cases, are far too various to be captured in an illuminating way in a single general account" (p. 312).

Because of this twofold incompleteness of theories, an additional source of knowledge is needed, namely "personal judgment." According to Michael Polanyi (1946, p. 17), to deal with discrepancies between theories and observational results, we cannot do without personal judgment:

> There is no proof of a proposition in natural science which cannot conceivably turn out to be incomplete, so also there is no refutation which cannot conceivably turn out to have been unfounded. There is a residue of personal judgement required in deciding—as the scientist eventually must—what weight to attach to any particular set of evidence in regard to the validity of a particular proposition.

According to Daston and Galison (2007, p. 325), mechanical objectivity alone, that is, a "rigid adherence to rules, procedures, and protocols," would not suffice; additional "trained judgment" is needed.

Thus, a calculus of observation to attain accurate measurement results includes the following elements: a calculation formula that is based on a representation of both measurand and the accompanying noise, mechanical procedures for calibration and attaining precision, theoretical assumptions about the laws governing the measurand, and trained judgment.

We will now trod carefully through the history of calculi of observations in order to explore the various meanings and shapes these elements have across time and across disciplinary boundaries, and we will pause whenever a relevant development occurs.

This history of calculi of observations includes—but is not equal to—a history of creating objectivity: the elimination of personal judgment by mechanical rules of calculation (see chapter 1). Likewise, Swijting (1987, p. 281) shows that the very introduction of formal standardized numerical methods such as the method of least squares was meant to have an objectifying effect, the elimination of personal judgment: "the strict use of numerical methods itself takes away a personal and thus arbitrary element from science. The existence of rules makes it easier for researchers not to be influenced by their preconceptions when they interpret their data." This strict use of numerical methods contributed to the general aim in science of achieving mechanical objectivity in measurement.

The goal of science to achieve full objectivity, however, was never and can never be fulfilled completely, since there always will be a "residue" for which evaluation, individual skill, and experience are indispensable.[8]

> The history of scientific objectivity is surprisingly short. It first emerged in the mid-nineteenth century and in a matter of decades became established not only as a scientific norm but also as a set of practices. . . . However dominant objectivity may have become in the sciences since *circa* 1860, it never had, and still does not have, the epistemological field to itself. . . . after the advent of objectivity came trained judgment. (Daston and Galison 2007, pp. 27–28)

3.2. Theory of Error

> We come, lastly, to the theory of errors. We are ignorant of what accidental errors are due to, and it is just because of this ignorance that we know they will obey Gauss's law. Such is the paradox. . . . We only need to know one thing—that the errors are very numerous, that they are very small, and that each of them can be equally well negative or positive. What is the curve of probability of each of them? We do not know, but

> only assume that it is symmetrical. We can then show that the resultant error will follow Gauss's law, and this resultant law is independent of the particular laws which we do not know. Here again the simplicity of the result actually owes its existence to the complication of the data.
>
> —Henri Poincaré, *Science and Method*, pp. 80–81[9]

The origin of the problem of accuracy can be found in Galileo Galilei's *Dialogue Concerning the Two Chief World Systems*, where he discussed methods of determining the position of a celestial body, the new star of 1572 (Hald 1986; Klein 1997, pp. 149, 151; and Maistrov 1974, pp. 30–34). Twelve observations were made, all of which gave conflicting positions. The problem was deciding which position was the correct one and was put forward by the character Simplicio:

> I should judge that all were fallacious either through some fault of the computer or some defect on the part of the observations. At best I might say that a single one, and no more, might be correct, but I should not know which one to choose. (Galileo [1632] 1967, p. 281)

Judy Klein (1997, p. 151) lists the conclusions at which the three characters, Salviati, Sagredo and Simplicio, in the *Dialogue* eventually arrive and which form the basic assumptions of a theory of errors upheld till today:

1. Errors are inevitable.

   There is "some error in every combination of these observations. This I believe to be absolutely unavoidable, for the observations used in every investigation being four in number (. . . made by different observers in different places and with different instruments), anybody who knows anything about matters will say that it cannot be that no error will have fallen in among the four. Especially when we know that in taking a single polar elevation with the same instrument in the same place and by the same observer (who may have made it many times), there will be a variance of a minute or so" (Galileo [1632] 1967, pp. 289–290).

2. There is no bias to overestimation or underestimation.

   "I do not doubt that they are equally prone to err in one direction and the other" (p. 291).

3. Small errors are more probable than larger ones.

   For "granted that these were wise and expert men, one must believe that they would be more likely to err little than much" (p. 290).

4. The size of the errors depends upon the precision of the instrument.[10]

Maistrov (1974, p. 33) mentions also a fifth conclusion:

5. The greatest number of measurements ought to be concentrated around the true value.

> "And among the possible places, the actual place must be believed to be that in which occur the greatest number of distances, calculated on the most exact observations" (Galileo [1632] 1967, p. 293).

Thus, Galileo arrived at the conclusion that errors in measurement are inevitable, that the errors are symmetrically distributed, that the probability of error increases with the decrease of the error size, that part of the error is due to the imprecision of the instrument, and that the majority of observations cluster around the true value (see also Maistrov 1974, p. 33).

Simplicio believed that the truth lay only in a single actual observation. This belief was common in Galileo's days, and had still to be argued against in the eighteenth century, when it became more and more accepted that the arithmetical mean of observations was an appropriate method to treat observational errors. For example, Klein quotes Thomas Simpson, who complained in 1755 (also quoted by Stigler 1986, p. 90)

> that the method practiced by astronomers, in order to diminish the errors arising from the imperfections of instruments, and of the organs of sense, by taking the Mean of several observations, has not been so generally received, but that some persons, of considerable note, have been of opinion, and even publicly maintained, that one single observation, taken with due care, was much to be relied on as the Mean of a great number. (Simpson 1755, pp. 82–83)

Instead of taking one observation with due care to reduce error, Simpson argued that

> the taking of the Mean of a number of observations, greatly diminishes the chances for all the smaller errors, and cuts off almost all possibility of any great one: which last consideration, alone, seem sufficient to recommend the use of the method, not only to astronomers, but to all others concerned in making of experiments of any kind. . . . And the more observations or experiments there are made, the less will the confusion be liable to err, provided they admit of being repeated under the same circumstances. (Simpson 1755, pp. 92–93)

In other words, this method of taking the mean of observations is based on the assumption that by averaging the observations, the observational errors will cancel out.

Thus, although the arithmetical mean became to be considered as a method to obtain true values, it was not clear why it should produce this result, that is, why the errors would cancel out. The reasons Galileo gave were not considered to be valid because they were connected to human agencies, in particular to "wise and expert men." A crucial step to the justification of this method was made by Carl Friedrich Gauss in his *Theoria Motus Corporum Coelestium* (*The Theory of the Motion of Heavily Bodies*), published in 1809. The main topic of this book is an investigation of the mathematics of planetary orbits. At the end, he added a section on the combination of observations. In this section he discusses the method of least squares. Although Gauss had invented the method of least squares, it was actually Adrien-Marie Legendre who introduced it to the world in an appendix to an astronomical memoir in 1805. Legendre stated the principle of least squares for combining observations and derived the equations from which the least squares estimates can be calculated. However, he provided no justification for the method, other than noting that it prevented extreme errors from prevailing by establishing a sort of equilibrium among all the errors (see Stewart 1995).

To arrive at his account, Gauss first discusses the nature of the errors. He distinguishes between two kinds of errors, "random or irregular errors" and "constant or regular errors," which distinction is based on the assumed sources of error.[11] A random error "depends on varying circumstances that seem to have no essential connection with the observation itself" (Gauss [1821] 1995, p. 3).

> Such errors come from the imperfections of our senses and random external causes, as when shimmering air disturbs our fine vision. Many defects in instruments, even the best, fall in this category; e.g., a roughness in the inner part of a level, a lack of absolute rigidity, etc. (Gauss [1821] 1995, p. 3)

These errors are the ones that were to be considered in his investigation; regular errors were explicitly excluded from it.

The innovative feature of the approach was that Gauss assumed that the possible values of errors ε have probabilities given by a function $\varphi(\varepsilon)$. Gauss noted that a priori he could only make general statements about this function. "In practice we will seldom, if ever, be able to determine φ a priori" (p. 7). The function is "zero outside the limits of possible errors while it is positive within those limits"; "we may assume that positive and negative errors of the same magnitude are equally likely, so that $\varphi(-x) = \varphi x$ [and] since small errors are more likely to occur than large ones, φx will generally be largest for $x = 0$ and will decrease continuously with increasing x" (p. 7). Instead of imposing further conditions directly, he assumed the conclusion.[12] He adopted as a postulate that when any number of equally good direct observations of an unknown magnitude are given, the most probable value is their arithmetic mean, and subsequently he proved that the distribution must have the form of what would later be called the Gaussian, or normal, curve:[13]

$$\varphi(\varepsilon) = \frac{h}{\sqrt{\pi}} e^{-h^2 \varepsilon^2},$$

for some positive constant h, where h came to be viewed as a measure of precision of observation. He then showed how in the more general situation this error distribution led to the method of least squares.

In his *History of Statistics*, Stephen Stigler (1986, pp. 141–142) notes that Gauss's argument was essentially both circular and a non sequitur:[14]

> In outline its three steps ran as follows: The arithmetic mean (a special, but major, case of the method of least squares) is only "most probable" if the errors are normally distributed; the arithmetic mean is "generally acknowledged" as an excellent way of combining observations so that errors may be taken as normally distributed (as if the scientific mind had already read Gauss); finally, the supposition that errors are normally distributed leads back to least squares.

It was Pierre-Simon Laplace who cut this circularity with the central limit theorem (Laplace 1809). Laplace showed how this theorem could provide a better rationale for Gauss's choice of $\varphi(\varepsilon)$ as an error curve: If the sums of errors are considered, then the limit theorem implies they should be approximately distributed as the normal curve $\varphi(\varepsilon)$. Note that in central limit theory, however, it is assumed that the circumstances in which the observations are made are identical and that the normal curve is a good approximation if the number of observations is very large.

Stigler (1986, p. 11) posits explicitly that: "The method of least squares was the dominant theme—the leitmotif—of nineteenth-century mathematical statistics. In several respects it was to statistics what the calculus had been to mathematics a century earlier." Ten years after Legendre's publication of the method of least squares, it had become the standard tool to deal with measurement errors in astronomy and geodesy in France, Italy, and Prussia. By 1825 the same was true in England. In the Netherlands, the method was introduced in astronomy in 1840 by the astronomer Frederik Kaiser. Typically the method was discussed in a book entitled *Eerste metingen met den mikrometer* (First measurements with the micrometer), treating the measurement results of a precision instrument (Dekker 1992). Both astronomy and geodesy were driven by this theory of error as applied to the choice of the best value to be adopted for the measurement of a physical quantity when there are a large number of independent determinations, equally trustworthy so far as skill and care concerned, yet differing from one another within the limits of actual measurement.

The method of least squares is a calculus of observations that consists of only one element, namely the arithmetical mean, as a mechanical procedure of attaining precision. This is sufficient for the purpose for which it was designed. The method of least squares is a statistical method to underpin the use of the arithmetical mean of measurements obtained with precise instruments. A science like astronomy concerned itself with the repeated measurement of physical quantities, like stellar positions, that were supposed to be without variation and where measurements were equally trustworthy.

These research conditions, however, do not apply to many other disciplines, as diverse as meteorology and actuarial science. Nineteenth-century meteorology concerned itself very little with the repeated measurement of physical quantities that were supposed to be without variation, and one could not rely on the assumption that the measurements were equally trustworthy. In actuarial science there are no instruments used at all; measurements are obtained by counting and calculation on the basis of different years, volumes, or periods. It came about that even at the end of the nineteenth century the method of taking the arithmetical mean of observations while applying the method of least squares was still the subject of discussion and considered to be less relevant in actuarial practices.

There was, however, also an important change in the goal of the application of the method of least squares when it traveled from astronomy and geodesy to other fields. According to Stigler (1986, p. 309), the tools developed "for *observations* in astronomy and geodesy, where a more or less objectively defined goal made it possible to quantify and perhaps remove nonmeasurement error," came to be applied "to social and economic *statistics*[,] where the goal was defined in terms of the measurements themselves and the size and character of the variation depended upon the classifications and subdivisions employed."

3.3. Measurement in Meteorology

The method of least squares is based on the theory of the properties of large numbers of observations of stable patterns, like a planetary orbit. However, in meteorology this method is used in an entirely different manner. It is applied to observations that are designed and intended to record change of patterns by measuring, for example, daily, monthly, seasonal or secular changes in the meteorological elements such as pressure, temperature, humidity, rainfall, and sunshine. It therefore borrows from statistics the methods originally developed to deal with *errors*, and applies them to deal with *variation* in variables like annual temperature or rainfall, the measurement of which assumes that the errors can be ignored. These variations are assumed to be the result of natural causes that are certainly real, but unknown to the observers. These variations are interpreted as deviations from a "normal." The idea of a normal for a period of years is that the mean value for any period of equal length in a very long series would give the "normal" value, and this would necessarily imply perfect recurrence at the completion of a year.

According to the Dutch polymath Buys Ballot, deviations from normal (i.e., these calculated means over long periods) say much more than direct measurements.[15] This method based on the deviations from normal led to his famous meteorological law connecting wind direction and location of areas of high and low pressure.[16] Moreover, and more related to the subject of this chapter, Buys Ballot had to deal with measurements by unreliable instruments collected from different places around the world. He learned that deviations from a normal were more trustworthy than the direct measurements by the instruments; the deviants eliminated the (unknown) deficiencies of these instruments, as will be explained below in more detail.

Ballot explicated his method to uncover the main causes of changes of temperature and pressure in a paper "On the Great Importance of *Deviations* from the Mean State of the Atmosphere for the Science of Meteorology" (1850). This method was based on four "propositions":

> I. The average temperature which prevails at any certain place is not that which is generated there by the action of the sun, &c., and which would depend simply on the latitude and the elevation of the ground, but is remarkably changed by the influences of other regions, particularly by the action of the winds.
> II. That average temperature, such as it is obtained anywhere from observations during a series of years, for the different months or days of the year, will by no means always prevail at those places for the determined month or day of each single year. On the contrary, observations give generally great variations; and it is precisely the magnitude of these variations which it is of the utmost importance to learn.
> III. What we asserted regarding the temperature in proposition II. applies equally to all meteorological indications; it is of great importance to become acquainted with those variations of the barometer, and of the force and direction of the wind.
> IV. The most efficient means for prognosticating the weather are, the employment of the electric telegraphs and of self-registering instruments, because they facilitate and make possible a tabular union of the variations mentioned in II. and III. (Buys Ballot 1850, p. 43)

Based on these propositions, he defined six different kinds of averages that would enable him to measure the causes of changes in temperature: Θ, the "mean theoretical temperature of a place," assumed to result over a long period of time (e.g., a year or a season); θ, the mean theoretical temperature of a place resulting from a shorter period of time (e.g., a month or a day); *MT*, the "mean temperature deduced from long series of observations," referring to a yearly or seasonal average; *mt*, the mean temperature deduced from long series of annual observations, referring to a monthly or daily average (in Buys Ballot's later publications referred to as "normal"). Last, there is *OT*, the "observed mean temperature," the average temperature of one particular year or season; and *ot*, the average temperature of one specific day or month.

The values of the theoretical temperatures Θ and θ of a particular place are assumed to be determined only by the latitude and altitude of that place, that is, the temperatures that would be measured if that place did not receive heat from and emit heat to the surrounding environment. Theoretically we should infer these values from

> the warmth which emanates from the sun to us every day; from the warmth which every day and night issues from beneath the surface of earth; and from the warmth that is produced by animals, consumed by plants, lost by radiation, given by condensation of vapour. (Buys Ballot 1850, p. 44)

MT and *mt* are the averages of series of temperature observations made at different locations at the same altitude and latitude. So *MT* and *mt* could be used as estimates of Θ and θ respectively. The value of, for example, *MT* so obtained would, however, be somewhat greater than Θ,

> because there is more air drawing towards the north than towards the south over a whole parallel circle, that southern air at the same time being warmer; and also because, near the equator, the latent heat which is employed in the vaporizing of water is greater than that which is freed by rain; and, on the contrary, in higher latitudes there is more freed than expended on the formation of vapour. (Buys Ballot 1850, p. 44)

Thus, the difference between Θ and *MT* should be ascribed to the influence of the wind during a season; similarly the difference between θ and *mt* should be ascribed to the influence of the wind during one month or one day.

MT should therefore be considered as the "equilibrium state of temperature at a determined season of the year at that place." The differences $OT - MT$ and $ot - mt$ should then be ascribed to circumstances such as the wind not having its usual direction in that period of time, or the distribution of temperature at the surrounding places being different from usual.

> Thus it is necessary that we know the variations, not only for the place itself for which we desire to explain the temperature, but also for the surrounding places, since the variations at the first place must be explained partly from the variations at the latter. *The most important causes are always to be sought in the variations* (deviations); it is from those that we must derive the exhibition of the state of temperature, not from the absolute observed temperature. (Buys Ballot 1850, p. 45; emphasis added)

These conclusions also applied to barometer readings: "the deviations again are the greatest importance, especially as here the theoretical state is known for every latitude" (p. 45).

The method Buys Ballot implicitly proposed in his paper "On the Great Importance of Deviations from the Mean State" was the "method of residues," as it was called by William Whewell ([1847] 1967). According to Whewell, this method should be applied when a combination of influences is operating at the same time:

> When we have, in a series of changes of a variable quantity, discovered *one* Law which the changes follow, detected its Argument, and determined its Magnitude so as to explain most clearly the course of observed facts, we may still find that the observed changes are not fully accounted for. When we compare the results of our Law with the observations, there may be a difference, or as we may term it, a *Residue*, still unexplained. But this Residue being thus detached from the rest, may be examined and scrutinized in the same manner as the whole observed quantity was treated at first: and we may in this way detect in *it* also a Law of change. If we can do this, we must accommodate this new found Law as nearly as possible to the Residue to which it belongs; and this being done, the difference of our Rule and of the Residue itself, forms a *Second Residue*. This Second Residue we may again bring under our consideration; and may perhaps in *it* also discover some Law of change by which its alterations may be in some measure accounted for. If this can be done, so as to account for a large portion of this Residue, the remaining unexplained part forms a *Third Residue*; and so on. (Whewell [1847] 1967, pp. 409–410)

When comparing this method with the method of means, that is, the method of least squares, Whewell notes that the two methods are opposites: "For the Method of Residues extricates Laws from their combination, *bringing them into view in succession*; while the Method of Means discovers each Law, not by bringing the others into view, but by *destroying their effect* through an accumulation of observations" (p. 411).

This observation was echoed by Buys Ballot in one of the 21 propositions he formulated as the scientific aims of meteorology.[17] The fourteenth proposition noted that "the averages cover all influences that do not manifest themselves each year on the same date; deviations make them conspicuous" (Buys Ballot 1851, p. 2; translated by the author).

Whewell referred to John Herschel's *Preliminary Discourse on the Study of Natural Philosophy* (1830) for a treatment of this method in a wider sense.

> Complicated phenomena, in which several causes concurring, opposing, or quite independent of each other, operate at once, so as to produce a compound effect, may be simplified by subducting the effect of all the known causes, as well as the nature of the case permits, either by deductive reasoning or by appeal to experience, and thus leaving, as it were, a *residual phenomenon* to be explained. (Herschel 1830, p. 156)

Herschel was England's leading scientist at that time. He had initiated a worldwide network for meteorological observations during his stay in South Africa (Cannon 1961). From 1835 onwards, meteorological observations were carried out on the twenty-first of the months March, June, September, and December at more than sixty locations around the world. Participants were expected to make these observations every hour for a period of 36 hours according to specific instructions, specified in his "Instructions for Making and Registering Meteorological Observations at Various Stations in Southern Africa and Other Countries in the South Seas, as Also at Sea" (1836). He terminated the project in 1838 because he considered the grid of this worldwide network to be too coarse (van Lunteren 1998, p. 219). However, he did ask the Belgian polymath Lambert Adolphe Jacques Quetelet to carry out this project on a smaller scale in Belgium. This Belgian network soon spread across the whole of Europe.

Herschel's instructions were meant as "the means of rendering their observations most available for useful purposes, and comparable with each other, and, with those intended to be referred to as standards" (Herschel 1836, p. 138). The instructions consisted of three parts: (1) "General Recommendations and Precautions," (2) "Of the Times of Observations and Registry," and by far the largest part, (3) "Of Meteorological Instruments, and First of the Barometer and Its Attached Thermometer." What is striking about these instructions is that they are very detailed, in the sense of that they stretch to every tiny detail of measurement readings. For example, observers were instructed: "Before reading off, give a few taps on the instrument, enough to make the upper end of the column of quicksilver *shake* visible, as the mercury is apt to adhere to the glass and give erroneous readings" (p. 141).

Buys Ballot found a worldwide network of observations such as Herschel's original one very valuable. His first experience with such a network was acquired during his student days, when he and his study friend Frederick Wilhem C. Krecke were involved in a project run by Richard van Rees, their mathematics and physics professor at the University of Utrecht (see van Everdingen 1953, p. 24, and van

Lunteren 1998, p. 226). For the years 1839 to 1843, the two students assisted van Rees in making meteorological observations on the twenty-first of the months March, June, September, and December (van Everdingen 1953, p. 27, and Buys Ballot 1882, p. 2).

Twelve years later, soon after the establishment of a meteorological observatory at Utrecht, Buys Ballot published an appeal to "all friends of meteorology" to participate in a project to provide observations from across the Netherlands (Buys Ballot 1848). Like Herschel, he realized the importance of instructions, so they were included in this appeal. But while Buys Ballot aimed at a same kind of network as Herschel's, his view on the nature of the instructions for the dispersed observers opposed those given by Herschel. Buys Ballot made it rather quite explicit that the observations need not be as precise as those in astronomy, and just three observations at "convenient" moments would be welcome; he would even be pleased with twice-daily observations. Moreover, the observations need not be carried out at the same time each day.

To underpin the ease of his instructions, Buys Ballot even quoted one of his opponents as a starting point for explicating them:

> For me it is impossible to suppress a feeling of distrust whenever I consider these immense series of observations, from which one hopes to attain knowledge of the laws and causes of the phenomena in our atmosphere. Nobody will assert that *these* fully meet the requirements of precision, and that no greater correctness is now achievable and necessary; but one should also note the relatively minor knowledge that is inferred from them, and the uncertainty to which this is still rightly subjected. (van der Willigen, quoted in Buys Ballot 1848, p. 380; translated by the author)

Volkert Simon Maarten van der Willigen replied to Buys Ballot's appeal with an extensive discussion of the sense and meaning of precision in the natural sciences (van Lunteren 1998, p. 231): "It cannot be said often enough: to the physicist precision and progress in science are the same thing (van der Willigen, quoted in van Lunteren 1998, p. 231; translated by the author).

Just a year before this exchange van der Willigen had received his Ph.D. under the supervision of the earlier-mentioned Leiden astronomer Kaiser, who was internationally renowned for his precise measurements (Beek 2004). This pursuit of precision was generally not uncommon in those days, because, particularly in astronomy, one's reputation as a scientist depended on it. In astronomy, the quality of the observations revealed the quality of the researcher, and therefore correctness and preciseness were the preconditions to a successful scientific career.

However, in meteorology—in Buys Ballot's perception—the number and spread of observations were more important than greater preciseness. In his propositions published in 1851, he explained why:

> After all, it is a good thing that very extensive observations are carried out at some observatories, which fill bulky volumes: although this is important for *Climatology*, it is preposterous to think that it meets the needs of *Meteorology*. The weather situation in one location depends on that in surrounding locations; thus observations in one location cannot teach us more than we are already likely to infer from the astute connection of barometer, thermometer and wind vane readings.
>
> . . . As linking long series observations in one location to those of the surrounding places opens an important new perspective, it is now infinitely better to conduct simple observations in one hundred locations not too far apart than to conduct very complete observations in just ten locations. (Buys Ballot 1851, p. 2; translation by the author)

Buys Ballot was well aware that to receive as many observations from as many locations as possible, he had to deploy "friends of meteorology" who were not always experts in using the best instruments available and who would not always be able to make their observations under optimal conditions. To study meteorology, quantity and spread were preferred above quality and precision. His remedy to deal with the imperfect observations was to confront the imperfect observation with the "normal," which determination is based on these imperfect observations. By taking the difference between these two values, the deviation, Buys Ballot assumed that the imperfection was eliminated, as will be explained below.

The readings of the thermometers and barometers were published in yearbooks. The first was published in 1851 and contained tables with the daily deviations from the normals for the years 1849 and 1850. The first yearbook contained only Dutch data, but subsequent yearbooks gradually included more and more measurements and deviations from normals in foreign locations. These data from other locations in Europe were used to calculate the normals for these locations.

Ultimately Buys Ballot's aim was to have a worldwide network of observatories:

> The whole of the globe must be covered with a network of observatories, where observers placed at equal distances are able to watch all phenomena of the weather as accurately as possible. . . . which is preferable, observations at a thousand well disposed observatories for two or three years in addition to more years at some of them, or a series of hundreds of years at a dozen stations? We prefer the former alternative. (Buys Ballot 1872, pp. 18–19)

Naturally, such a network required international coordination and standardization of the observations. The first international meeting, in Leipzig, was held in 1872. This meeting was considered a preparatory meeting for the official conference to be held in Vienna the following year. In preparation for the Leipzig meeting, Buys Ballot wrote the booklet *Suggestions on a Uniform System of Meteorological Observations* (1872). His aim of a global system exacerbated the problem that from all over the world would arrive observations the validity of which was by no means certain. Therefore, Buys Ballot insisted once again on publishing the deviations from the normals: "Departures are perfectly independent of the daily and annual range, and of the local disposition and correction of the instruments, because the normals are likewise computed for these circumstances" (p. 18).

The problem is that if a barometer has been moved or replaced, or the time of observation has changed, it is as "if the climate of a place is changed" (Buys Ballot 1872, pp. 14–15; 1882, p. 51). For this reason, Buys Ballot persistently chose to publish departures from the normals: comparing a reading of a certain instrument with the normal reading of that instrument removes the effects of the errors of the instrument and the effects of the latitude (Buys Ballot 1872, p. 16; 1882, p. 52). So Buys Ballot in fact suggested to the international community that recorders of data should import what he called the "Dutch system" of presenting observations. This "Dutch system" was described in a paper published in an engineering journal:

> It is obvious, but not generally acknowledged, that no absolute reading of the barometer has any significance, but only the difference (called departure) of an actual reading with the average reading of *that* instrument at the *same* place, at that latitude, longitude and height above the sea on the same day. The departure is the true and accurate measure of the perturbance [*sic*],[18] and intimately connected, but, as we shall see, not identical, with the force that tries to restore the equilibrium. The single reading of the barometer, on the contrary, is an arbitrary number of no signification at all, unless you substitute an accurate approximation of the average height of the readings. (Buys Ballot 1865, p. 246)

This "Dutch system" came later to be known as "Buys Ballot's principle" (van Everdingen 1953, pp. 86–87).[19] According to "Buys Ballot's principle," it was better to work with deviations than with absolute observations. Deviations have the advantage that they eliminate the errors of the instrument and position (see also KNMI 1954, p. 19). If a barometer has the wrong zero point or is moved to a different altitude than assumed, the difference between an observation and a normal would nullify these instrumental errors. "Each location itself should ensure the reliability of its instruments. The comparisons reveal all their flaws" (Buys Ballot quoted in KNMI 1954, p. 18; translated by the author).

Buys Ballot's calculus of observations can be reconstructed as follows. The facts to be established are the daily variations in, for example, pressure. They have to be inferred from daily readings of unreliable instruments. To deal with these instruments' errors, the method of the arithmetical mean—normal—is used, not to erase them but, on the contrary, to capture them. The mean, which includes the instrument's error, is compared with the measured observation, which also includes the same instrument's error. By taking the difference between these two values, the deviation, the instrument's error is eliminated.

To present it more formally, let y_t be a reading of an instrument supposed to measure, say, pressure at a specific moment, t, on a specific day, for example, 21 March at noon. Each particular moment on a particular day, t, is characterized by a specific value, x_t, indicating the "normal" pressure, typically for that particular moment on that particular day of the year. The instrument's error ε is caused by unknown defects of the instrument itself (not calibrated or calibrated with the wrong standards, e.g., the barometer is placed at the wrong altitude). Buys Ballot was interested in the daily variations z_t. As a result:

$$y_t = x_t + \varepsilon + z_t.$$

To reveal the deviations z_t, Buys Ballot first calculated the normal for that particular moment. A normal N_t was an arithmetical mean of all measurement readings of that particular instrument at the same moment for as long a period as possible, for example, 10 years. He assumed that this normal would consist of the unknown true value and the unknown instrument's error:

$$N_t = x_t + \varepsilon.$$

Then the deviation z_t, obtained by subtracting the normal from the observation,

$$z_t = y_t - N_t,$$

no longer includes the instrument's error ε. This reconstruction shows clearly how Buys Ballot used the arithmetical mean, not to neutralize measurement errors, but instead to capture them.

Buys Ballot's principle was a calculus of observations that did not consist of a procedure for attaining precision. Its two main elements were a mechanical procedure for calibration (the comparison of the observation with the normal that functions as a standard), and the theoretical assumption that there exists a mean theoretical temperature (or pressure) that prevails at any specific location.

3.4. Actuarial Science

In actuarial science, where the available data are often in the form of "time series," not every observation has the same significance to determine the true value of, for example, the birth rate or mortality rate at a certain moment in time. The same problem occurs when "graduation" (that is, smoothing of the time series to eliminate errors) is needed in the population distribution of age or income. Instead of simply taking an arithmetical mean, which assumes the equal significance of each observation, the true values should be determined by a weighted average of a number of years or categories, or by weighted volumes.

To explore this problem, we first discuss the work of the Dutch actuary Corneille Landré.[20] As a student, Landré attended courses given by Buys Ballot and, for a while, worked at the Meteorological Institute headed by Buys Ballot, where he calculated the normals of the barometers of various observatories. He may therefore have inherited his critical attitude from Buys Ballot, because he shared Buys Ballot's critical view of the applicability of the arithmetical mean as the standard method for treating measurement errors. Nevertheless, Landré developed his own calculus of observations adapted to the nature of observations in actuarial science.

One of Landré's involvements in actuarial science was the construction of mortality (or life) tables. One important aspect of the construction of such tables is the graduation—smoothing—of the mortality data. The irregularity of a time series is considered to be an indication of errors of measurements, which should be removed. But in the case of mortality, only one measurement of the phenomenon is possible for each moment of time. Measurements taken at other moments (past and future) are used to remove errors, but this cannot be done by simply averaging these measurements.

Landré had always been very critical of the application of the method of least squares to actuarial science, as it was based on "hazardous suppositions" and would not always lead to the best solution (Mounier 1906b, p. 309). Landré never published his doubts about the applicability of the method of least squares to actuarial science (although he did write two articles on the method (Landré 1881, 1884), but his fellow editor of *Archief voor de Verzekerings-Wetenschap en Aanverwante Vakken* (Archives of Actuarial Science and Related Subjects), Guillaume Jacques Daniel Mounier, wrote two articles on this method at Landré's suggestion (Mounier 1903, 1906b).

Mounier criticized the method of least squares as an appropriate method for actuarial science by explicating the four "suppositions" of its applicability:

1. The probability of error $+\varepsilon$ is equal to the probability of error $-\varepsilon$.
2. The best method and the best instruments are used as well as possible.
3. The number of observations is very large, preferably as large as possible.
4. The probability of the error is an infinitely small function of the error itself. (Mounier 1903, p. 8; translated by the author)

According to Mounier these suppositions reveal the original intended application of this method, namely, inference of the most probable value of a variable from a long series of precision measurements giving inevitably different results. To make this method applicable to actuarial science, the suppositions should be revised by replacing the term "error" by "deviation" and by removing any reference to instruments. In the subsequent discussion of these four suppositions, he argued that supposition 1 follows from supposition 2 and that the latter means that the observations should be as precise as possible. The third supposition is related to the law of large numbers and aims at eliminating all random influences. In Mounier's view it means that the method of least squares does not apply when the number of observations is not large.

The general approach to deal with observational errors in actuarial science is graduation or smoothing, which Morton D. Miller (1946, p. 4) defined as "the process of securing from an irregular series of observed values of a continuous variable a smooth regular series of values consistent in a general way with the observed series of values."[21] Graduation is based on the view that there is an underlying "law" that produces a smooth, regular, and continuous sequence of values, but that all kinds of disturbances have turned this sequence into an irregular one. The irregularity represents deviations from the true values, and thus the revised, "graduated," sequence should be taken as a representation of the underlying law. However, the only empirical knowledge about these laws we have are these observations.

> Since no law of mortality, in the sense of a physical law, is known to us, nor is one likely to be discovered, we have no way of knowing *a priori* what the basic pattern of mortality is. We must therefore rely on the information supplied by observations of the rates of mortality actually being experienced. (Miller 1946, pp. 1–2)

But for the analysis of these observations there are too many possible methods of graduation available to choose from; the choice of the most appropriate graduation method is underdetermined by the observations. To come to a definite choice additional assumptions are needed, such as an assumption about the assumed underlying law, but, as Miller noted: "The theoretical reasons justifying graduations are those upon which this assumption rests" (p. 5). Therefore it is too much to expect that the errors in the observations will be completely eliminated. There will always be a "residual error." For that reason Miller suggests that we think of the graduated series as a "representation of the underlying law rather than as the law itself" (p. 5), so we are aware that in principle various different representations are possible.[22]

Miller discusses five different methods, the "graphic method," the "interpolation method," the "adjusted-average method," the "difference-equation method," and "graduation by mathematical formula," of which only the "adjusted-average method" involves mean squared error considerations. This latter method is usually referred to as the moving-weighted-average method, because it raises the problem of choosing the appropriate adjustment, that is, weighting system.

It is worth pausing here to have a closer look at Landré's discussion and comparison of various kinds of these moving-weighted-average graduation methods in several of his publications on this subject (Landré 1889, 1900a, 1900b).

To find the true value x_t of a particular property of a phenomenon to be investigated, for example the number of deceased people at age 40, a graduated value $\hat{x}_t$ is produced as a weighted average of a certain number (n) of consecutive observed values. Without loss of generality we will restrict the discussion to time series; hence t is a time index. The basic formula is

$$\hat{x}_t = \sum_{i=-n}^{n} a_i y_{t+i},$$

where the a_is constitute the weighting system and y_t the observed time series. In most cases the moving-weighted-average formulas are symmetric: $a_i = a_{-i}$. The problem is to find interpretations of a_i such that $\hat{x}_t$ is the best estimate of x_t.

The general rationale behind most of the graduation methods is to take as starting point the observation equation (cf. eq. 3.1):

$$y_t = x_t + \varepsilon_t.$$

Substituting this observation equation into the basic graduation formula, we obtain

$$\hat{x}_t = \sum_{i=-n}^{n} a_i x_{t+i} + \sum_{i=-n}^{n} a_i \varepsilon_{t+i}.$$

The problem is to find an interpretation of a_i such that x_t is reproduced, that is,

$$x_t = \sum_{i=-n}^{n} a_i x_{t+i},$$

and such that the measurement error in $\hat{x}_t$, which is

$$\hat{\varepsilon}_t = \sum_{i=-n}^{n} a_i \varepsilon_{t+i},$$

is substantially smaller than ε_t, the observational error.

To find these values, it is assumed a priori that the sequence x_t (representing a law or regularity) is closely represented by a smooth function, like a low-degree polynomial such as a cubic, over a limited range $[t - n, t + n]$. If y_t is unbiased, then $\hat{x}_t$, given by the moving weighted average, is also unbiased (if the cubic assumption holds), so that $E[\hat{\varepsilon}_t] = 0$. Thus, under the assumption of unbiased observations, there is no merit in trying to minimize expected measurement error. Instead, the expected squared measurement error, $E[(\hat{\varepsilon}_t)^2]$, which is the variance of the measurement error, $Var(\hat{\varepsilon}_t)$, is minimized. And since $\hat{x}_t = x_t + \hat{\varepsilon}_t$, it follows that $Var(\hat{\varepsilon}_t) = Var(\hat{x}_t)$.

But so far this moving-weighted-average approach is still too unconstrained; an infinite number of graduation formulas is still possible. Landré discusses several options, not only those of Finlaison, Woolhouse, Higham, and Karup (based on different smooth functions), but also combinations of them. To evaluate these different formulae, Landré used the following additional criterion, which he called a "new principle": the revised (graduated) value $\hat{x}_t$ of y_t should be a *correction* of y_t and not a *rejection*:

$$\hat{x}_t = y_t + k_t.$$

That is, the correction term k_t should be such that it does not remove the original observation, y_t, from the formula: a_0 should not be equal to zero. Moreover, Landré explicitly preferred formulas for which the observation, y_t, has the greatest weight; in other words, the weight a_0 is larger than any other weight a_i (Landré 1900a, p. 326).

However, this criterion is not sufficient to choose among the still many remaining graduation formulas. The second criterion he proposed is to choose the smallest mean squared error—which is the above criterion of minimizing expected squared measurement error on which modern graduation methods are based. Landré's "new principle" as outlined above never became an accepted and used principle in actuarial science.

Landré's calculus of observations consists of the two mechanical procedures of calibration and attaining precision. For the calibration, however, Landré used the observation itself as a standard. The reason for this proposal was that he wanted to reduce the involvement of a priori theoretical assumptions as much as possible; rather, he aimed at graduation without having to assume a specific smooth function.

The development of graduation methods went in another direction, leading to kinds of calculi other than the one Landré had proposed. Instead of assumptions about the observations themselves, the focus shifted to assumptions on the observational errors. To derive sufficient conditions to determine the a_is—beside the conditions inferred from the assumed smooth function—the following additional assumptions are usually made: If we assume that the observed y_t differed from the underlying true value x_t by small errors, ε_{t+i}, "of an accidental nature" (De Forest quoted in Stigler 1978, p. 255), that is, the errors are uncorrelated with equal variances, say σ^2, then we can write

$$Var(\hat{x}_t) = \sigma^2 \sum_{i=-n}^{n} a_i^2.$$

Defining

$$R_0^2 = \sum_{i=-n}^{n} a_i^2,$$

we can now say that the coefficients a_i we seek for the graduation formula are those that minimize R_0^2 and satisfy the constraints given by the smooth function. A moving-weighted-average formula with coefficients so derived is called a Minimum-R_0 formula. Since $Var(y_t) = Var(\varepsilon_t) = \sigma^2$, it then follows that

$$R_0^2 = \frac{Var(\hat{x}_t)}{Var(y_t)}.$$

In other words, R_0^2 is a ratio of variances.

This approach is based on the works of Erastus Lyman De Forest.[23] However, De Forest discovered that the minimum square error criterion need not produce a very smooth relation globally, notwithstanding the assumption that the smooth function was, say, cubic locally. As an alternative to the preceding approach he proposed a criterion based on smoothness: minimize the probable error of the fourth difference of the smoothed series $\hat{x}_t$, or equivalently, minimize $E(\Delta^4 \hat{x}_t)^2$, subject to some smooth function constraint.

> As a contribution to nineteenth century work on smoothing or adjustment, De Forest's introduction of this measure of smoothness as an optimality criterion was well ahead of its time, and his work was not generally appreciated until [it was] rediscovered . . . in the 1920's. By then, others had come upon his main techniques independently. (Stigler 1978, p. 256)[24]

Thus, because graduations using the Minimum-R_0 formula "seldom prove to be satisfactory, frequently resulting in neither good fit nor an acceptable degree of smoothness" (London 1985, p. 38), a generalization of the Minimum-R_0 formula is preferred. The ratio being minimized is

$$R_z^2 = \frac{Var(\Delta^z \hat{x}_t)}{Var(\Delta^z y_t)}.$$

As in the $z = 0$ case, one can derive that

$$Var(\Delta^z \hat{x}_t) = \sigma^2 \sum\nolimits_{i=-n-z}^{n} (\Delta^z a_i)^2.$$

The uncorrelated and equivariance assumption leads to

$$Var(\Delta^z y_t) = \binom{2z}{z} \sigma^2.$$ [25]

Thus

$$R_z^2 = \frac{\Sigma(\Delta^z a_i)^2}{\binom{2z}{z}}.$$

The R_z^2 is interpreted as the rate of roughness in $\hat{x}_t$ to that in y_t, as measured in the zth order of differences (see London 1985, p. 41). If we want $\hat{x}_t$ to be smoother than y_t, we prefer to find $R_z^2 < 1$. R_z^2 is therefore called the "smoothing coefficient" of the moving-weighted-average formula.

In line with De Forest's approach, the moving-weighted-average calculus of observations came to consist of a rather sophisticated procedure for attaining precision that does not require an assumption specifying the smoothing function. The necessary assumptions are a theoretical assumption that the underlying law is smooth and assumptions about the properties of the errors. Because unbiasedness is assumed, it does not include a separate procedure of calibration. The standard in this calculus is the value for z, which establishes the degree of polynomial $(z - 1)$. So inherently a specific degree of a polynomial is used as a standard of smoothness. As a result, in De Forest's approach, precision and calibration are two sides of the same coin.

But graduation is not only characterized by "smoothness." The other "essential quality," according to Miller (1946, p. 5), is "fit, or consistency with the observed data." But these two different qualities of smoothness and fit are "basically inconsistent." Improving one is at the cost of the other. Therefore, "any graduated series must of necessity follow a middle course between optimum fit and optimum smoothness; it must represent the result of a compromise between the two" (p. 5). There exists, however, no standard for this "compromise," and therefore must be left to the judgment of the graduator: "a graduation method must allow the graduator some latitude in choosing the relative emphasis to place on smoothness and fit in the graduated series" (p. 5).

It was Whittaker (1922) who developed a "new method of graduation" that captured both qualities. To do this he formulated the problem as one that "belongs essentially to the mathematical theory of Probability" (p. 64). His graduation method was designed to obtain the "most probable" values of x.[26]

Whittaker proposed that one could express an a priori probability of a sequence of true values of x in terms of the measure of smoothness, because "before the observations have been made we have nothing to guide us as to the probability of this . . . except the degree of smoothness of the sequence" (p. 64). He defined this measure of smoothness as

$$S = \sum_t (\Delta^z \hat{x}_t)^2.$$

He then assumed that a sequence $\hat{x}_t$ that produces a smaller value of S is a priori more probable than one that produces a larger value: "there is an antecedent probability that if the observations had been more accurate the curve would have been smooth" (p. 64). So the a priori probability is

$$f_{\hat{x}}(\hat{x}_t) = c_1 e^{-\mu S}.$$

Next, he defined a measure of fit, F, which he called the "fidelity of the graduated to the ungraduated values" (p. 65), as

$$F = \sum_t (\hat{x}_t - y_t)^2,$$

and postulated "the normal law of error" (p. 65), that is, the a priori probability of the errors, $y - \hat{x}$, given the values $\hat{x}$, to be

$$f_{Y|\hat{X}}(y_t|\hat{x}_t) = c_2 e^{-F}.$$

Using the "fundamental theorem in the theory of Inductive Probability" (p. 65), that is, Bayes' theorem, Whittaker arrived at the a posteriori probability[27]

$$f_{\hat{X}|Y}(\hat{x}_t|y_t) = \frac{c_1 c_2 e^{-\mu S - F}}{f_Y(y_t)}.$$

The most probable value of $\hat{x}$ is that for which this a posteriori probability is a maximum, that is, for which $\mu S + F$ is a minimum.[28]

According to Whittaker this method has several advantages. One is its "elasticity" due to the freedom of choice of μ:

> A satisfactory method of graduation *ought* to possess such elasticity, because the degree to which we are justified in sacrificing fidelity in order to obtain smoothness varies greatly from one problem to another. (Whittaker 1922, p. 73)

So Whittaker's calculus of observations contains both procedures of attaining precision (fit) and calibration (smoothness), and a standard of smoothness (the choice of the value z), but now the theoretical assumptions are replaced by a "latitude" for the graduator in choosing the relative emphasis to place on smoothness and fit.

As Miller (1946) emphasized in his textbook, "graduation does not have a single solution" (p. 7). It depends upon the choice of the method, upon a choice on how much fit and how much smoothness, upon the field of application, but also "upon the skill and experience of the graduator" (p. 7). And, moreover the choice of the method will rest upon:

(i) The purpose to which the graduated table is to be put;
(ii) The form in which the data are given and their general characteristics;

(iii) The extent of the data; and
(iv) The experience, technical knowledge and preferences of the graduator.

> These factors, of course, are not fully separable but rather interdependent. (Miller 1946, p. 54)

Although Miller reassures us that as soon as an appropriate method has been chosen, "the arithmetical work may be handled by a computer without technical knowledge" (p. 55), a calculus of observations is never a completely mechanically objective methodology. There is always an experienced "graduator" needed to make a decision, a "trained judgment" on the relative weight of each of the components of the measurement error: how much should one aim at precision, or calibration, or standardization. The reason for the need of this expert judgment is that these three epistemic "values" are conflicting. More of one will be at the cost of the other two (today one would call this a three-player zero-sum game). And there does not exist an overall objective cost function that could be used to enable an objective decision.

3.5. Conclusions

The history of measurement is often told as a history of creating objectivity by increasingly replacing personal judgment by mechanical rules of calculation and theory. This chapter has, however, shown that the various attempts to develop such an objective calculus of observations were unable to arrive at the ideal of complete objectivity, that there always remains an inevitable and indispensable "latitude" of personal judgment. The history of calculi of observations includes histories of calculi of observational errors. These calculi consist of mechanical procedures and theoretical assumptions, but these tools for objectivity are never sufficient to deal with errors completely. A residual of personal judgment is needed to complete a calculus of errors, and so a calculus of observations.

Daston and Galison's history of objectivity, connected to the history of the making of scientific images, contains a similar message: Although trained judgment came after the advent of objectivity, as they argue, "the new did not always edge out the old." The history of the making of scientific images is actually a history of three epistemic virtues: "truth to nature," "mechanical objectivity" and "trained judgment," which were the dominant virtues in more or less subsequent periods.

> Some disciplines were won over quickly to the newest epistemic virtue, while others persevered in their allegiance to older ones. The relationship among epistemic virtues may be one of quiet compatibility, or it may be one of rivalry and conflict. In some cases, it is possible to pursue several

> simultaneously; in others, scientists must choose between truth and objectivity, or between objectivity and judgment. Contradictions arise. (Daston and Galison 2007, p. 28)

The history of the calculus of observations as told in this chapter is not so much a history of epistemic virtues as it is a comparison of various calculi of observations that have existed in the past and exist now. The epistemological lesson that can be drawn from this comparison is that—irrespective of what the preferred epistemic virtue in some period was and is—both mechanical objectivity and trained judgment are indispensable elements of a calculus of observations. On the moral level they may conflict, but on the practical level they have to be integrated and balanced toward each other; a kind of consensus has to be reached.

One of the main reasons for this irreducible latitude of personal judgment is that a model is underdetermined by the observations. Chapter 5 will discuss the consequences of this for scientific judgment. To reduce subjectivity in science as much as possible, judgments are evaluated by standards of rationality. In other words, knowledge based on rational judgments is considered to be objective. But, as will be shown, the concept of rationality itself is also based on a model, and for choosing the most appropriate model "considered judgment" is needed.

Notes

1. Quoted in Whittaker and Robinson 1924, p. 179. The original quote is from Poincaré's (1912) *Calcul des Probabilités*, p. 171.
2. To keep the discussion even more simple, I disregard the current discussions in metrology that led to the GUM approach and discuss accuracy according to the classical error approach (see chapter 2).
3. See Warwick (1995, p. 339) for a description: "The room contained a wide variety of instruments for calculation."
4. Accuracy is defined as "closeness of agreement between a measured quantity value and a true quantity value of a measurand" (JCGM 200 2012, p. 21); see also chapter 2.
5. Precision is defined as "closeness of agreement between indications or measured quantity values obtained by replicate measurements on the same or similar objects under specified conditions" (JCGM 200 2012, p. 22).
6. This is comparable with Morgenstern's discussion, in chapter 1, of the similarity between "planned observation" and "experimentation."
7. See also the discussion in chapter 4 about Koopmans's strong apriorism required for measurement.
8. The evaluation of this "residual" is part of a Type B uncertainty evaluation, as discussed in chapter 2.
9. "There is no general theory of error, and there will never be one" (Boumans and Hon 2014, p. 1), but a more general framework must be possible. This shared insight by Giora Hon and me was the beginning a joint project that culminated in a volume *Error and Uncertainty in Scientific Practice*, edited by us and Arthur Petersen. See Hon 1989 for an excellent epistemological typology of errors.
10. Maistrov 1974 discusses a similar conclusion: "the size of the instrumental errors, so to speak, must not be reckoned from the outcome of the calculation, but according to the number of

degrees and minutes actually counted on the instrument" (Galileo [1632] 1967, p. 293). This, however, does not refer to the precision of an instrument but to the sensitivity of it.

11. In the 1820s, Gauss returned to the least squares in two memoirs, the first in two parts, published by the Royal Society of Gottingen under the common title "Theoria Combinationis Observationum Erroribus Minimis Obnoxiae," translated by G. W. Steward (1995) in "Theory of the Combination of Observations: Least Subject to Errors." The quotations are from this latter work. They refer to the first part ("Pars prior"), first published in 1821.
12. See Stigler 1986 for a detailed exposition of this argument.
13. In the case of observations, y_i, of only one value, x, the squared errors $\sum_i \varepsilon_i^2 = \sum_i (y_i - x)^2$ is "least" (minimal) when $1/n \sum_i y_i = x$.
14. Stewart (1995, p. ix) notes that the assumption of normality at that time was considered to be tenuous, and that Gauss himself later rejected the approach.
15. Christophorus Henricus Didericus Buys Ballot (1817–1890) was the founder and first director of the Royal Dutch Meteorological Institute. In 1873, he was appointed as the first chairman of the International Meteorological Committee, a precursor of the World Meteorological Organization.
16. In the northern hemisphere, if one stands with one's back to the wind, the area of low pressure is to the left. In the southern hemisphere the reverse is true.
17. These were presented in the introduction of the first yearbook of the Dutch meteorological institute (Buys Ballot 1851).
18. He probably meant perturbation, in the sense of causing deviation of motion or of other behaviour.
19. Ewoud van Everdingen is Buys Ballot's biographer and fourth director of the Royal Dutch Meteorological Institute from 1905 to 1938.
20. Corneille L. Landré (1838–1905) was the first person in the Netherlands to be appointed in 1895 by a life insurance and annuity company to a newly created position of "actuary." He was cofounder and later board member of the Vereeniging van Wiskundige Adviseurs (Society of Mathematical Consultants), founded in 1888. The last 10 years of his life he was editor of the *Archief voor de Verzekerings-Wetenschap en Aanverwante Vakken*. In his day, he was the actuary with the greatest international reputation in the Netherlands. These biographic details are to be found in three obituaries written by his daughter Henriette F. Landré (1906), G. J. D. Mounier (1906a) and M. C. Paraira (1905–1907).
21. The Actuarial Society of America has a long history on promoting textbook publications on graduation. The first such text was prepared by Robert Henderson and published in 1918, followed by a second edition in 1938: *Mathematical Theory of Graduation*. Henderson's work was superseded in 1946 by Miller's *Elements of Graduation*. This text served as an education reference for a number of years following 1950 (Nesbitt 1989, p. 622). Dick London's *Graduation: The Revision of Estimates* succeeded Miller's in 1985.
22. See section 5.5 of chapter 5 for a more extensive discussion of the possibility and consequences of having more than one representation.
23. This brief account of De Forest's approach is from Stigler 1978.
24. See also Miller 1946, p. 25.
25. $\binom{n}{m} = \frac{n!}{m!(n-m)!}$ and $n! = n(n-1)(n-2)\ldots 1$.
26. This method was further developed by Henderson in his 1924 paper "A New Method of Graduation" and his 1925 "Further Remarks on Graduation." But these contributions were merely an improvement of the minimizing procedure inherent to this method of graduation. Nevertheless this method became to be known as the Whittaker-Henderson method.
27. $f_{\hat{X}|Y}(\hat{x}_t|y_t) = \frac{f_{Y|\hat{X}}(y_t|\hat{x}_t)f_{\hat{X}}(\hat{x}_t)}{f_Y(y_t)}$.
28. In macroeconomics the minimum of $\mu S + F$ is known as the Hodrick-Prescott filter to separate the cyclical component of a time series from the unfiltered data.

4

The Problem of Passive Observation

> Haavelmo remarks that physicists are very clever. They confine their predictions to the outcomes of their experiments. They do not try to predict the course of a rock in the mountains and trace the development of the avalanche. It is only the crazy econometrician who tries to do that, he says.
>
> —Nancy Cartwright, *The Dappled World* (1999, p. 46)

4.1. Introduction

What is the epistemological scope of science outside the laboratory? What can field scientists know? This chapter will study the responses to this question that were given in economics over a period of 200 years. Most prominently the ideas of Trygve Haavelmo will be discussed, who, to my knowledge, gave the most sophisticated answer to the problem of what kind of knowledge can we acquire by passive observation only. Passive observation is understood here to be observation outside the laboratory, that is, observation of an object, process, or phenomenon without intervening with the subject of study or controlling its environment.

To make the study of this problem of passive observation feasible, the aim of science will be narrowed to the following (too) simple goal: the search and subsequent testing of stable patterns that can be used for explanation and prediction. These stable patterns are usually called "laws of nature." But where do we find these patterns? Current philosophers of science, like Nancy Cartwright (1999), believe that these patterns only occur where circumstances are similar, that is, in laboratories or in ceteris paribus environments.

If these philosophers were right, the possibility of science outside the laboratory would be problematic, particularly the possibility of economics as a science. Haavelmo believed the "crazy" opposite, expressed by the preceding remarks to Cartwright (1999, p. 46).[1] His remarks have to be seen as part of a longer discussion, started in the nineteenth century, whether and in what sense economics can be considered as an "exact science."[2] The pessimistic view is that economics can never become an exact science; the optimistic view is that it eventually may be possible.

This discussion, "that there is, or may be, a science" of economics, started with John Stuart Mill's *A System of Logic* ([1843] 1911). Mill did not regard economics as an "exact science" (p. 553).[3] The basic premises of economics state accurately how specific causal factors operate, but, according to Mill, they are statements of tendencies and are inexact rather than universal generalizations. Economists know the major causes of economic phenomena, but there are many "interferences" or "disturbing causes" (see Hausman 1992, p. 124).

In an exact science explanation is complete, that is, according to Mill, when the phenomenon to be explained (the explanandum) has been brought under laws comprehending the whole of the causes by which the phenomenon is influenced, "whether in a great or only in a trifling degree, whether in all or only in some cases, and assigning to each of those causes the share of effect which really belongs to it" (Mill [1843] 1911, p. 553). So a science is exact when "the greater causes, those on which the principal part of the phenomenon depends," Mill writes, "are within the reach of observation and measurement; so that if no other causes intervened, a complete explanation could be given not only of the phenomenon in general, but of all the variations and modifications which it admits of" (p. 552). In other words, an exact science is characterized by providing complete explanations. An explanation is complete when it includes a complete list of all the causal factors, whether great or small, whether always or sometimes, that have an influence on the explanandum.

In opposition to this, in an inexact science

> the only laws as yet accurately ascertained are those of the causes which affect the phenomenon in all cases, and in considerable degree; while others which affect it in some cases only, or, if in all, only in a slight degree, have not been sufficiently ascertained and studied to enable us to lay down their laws, still less to deduce the completed laws of the phenomenon, by compounding the effects of the greater with those of the minor causes. (Mill [1843] 1911, p. 553)

Mill refers to the science of tides as an example of an inexact science. Scientists know the laws of the great causes, the gravitational attraction of the sun and the moon, but they are ignorant of "circumstances of a local or casual nature, such as the configuration of the bottom of the ocean, the degree of confinement from shores, the direction of the wind, &c" (p. 553).

Since in Mill's view, economists know only the laws of the "greater causes" of the phenomena, they are unable to infer invariably and precisely what actually occurs. Economics is in this way an inexact science. The inability to infer what actually happens is a consequence of inexact theories, not merely of faulty data or mathematical limitations.

About 40 years later, Alfred Marshall made the same comparison of economics with the science of tides, to discuss economics as an inexact science:

> The laws of economics are to be compared with the laws of the tides, rather than with the simple and exact law of gravitation. For the actions of men are so various and uncertain, that the best statement of tendencies, which we can make in a science of human conduct, must needs be inexact and faulty. (Marshall [1890] 1920, p. 32)

Marshall devoted chapter 3 (of book 1) of the *Principles* to a discussion of the nature of "economic generalizations or laws." Why, he asked, should the "laws of economics" be less predictable and precise in their workings than the law of gravitation? The key to Marshall's view, according to John Sutton (2000, p. 4) lies in his claim that economic mechanisms work out their influences against a messy background of complicated factors, so that the most we can expect of economic analysis is that it captures the "tendencies" induced by changes in this or that factor.

> No one knows enough about the weather to be able to say beforehand how it will act. A heavy downpour of rain in the upper Thames valley, or a strong north-east wind in the German Ocean, may make the tides at London Bridge differ a great deal from what had been expected. (Marshall [1890] 1920, p. 32)

The discussion of precisely this central problem of what kind of science economics is was recently revived by Sutton (2000) in his Gaston Eyskens Lecture with the telling title "Marshall's Tendencies: What Can Economists Know?" In this lecture, Sutton argued that "if the analogy of the tides were valid in economics, life would be much easier for economists" (p. 5). The reason he gave for this is that in the early 1950s, a successful program, "the standard paradigm of applied economics," was designed and it worked perfectly for phenomena like tides. "If Marshall's analogy were valid, we would have seen spectacular progress in economics over the past fifty years" (p. 5). According to Sutton, this standard paradigm is the econometric program set out by Haavelmo in his "article" titled *The Probability Approach in Econometrics*, published as a supplement to the July 1944 issue of *Econometrica*.[4]

Sutton sees this "standard paradigm" (a science of the tides) as consisting of three properties:

> 1. The true model captures a "complete" set of factors that exert large and systematic influences,
> 2. All remaining influences can be treated as a noise component that can be modeled as a draw from some probability distribution, and
> 3. The model determines a unique equilibrium. (Sutton 2000, p. 20)

Although this paradigm, according to Sutton, works successfully for tide phenomena, it has failed for economic phenomena: Suppose there is a true model linking an endogenous variable y to a set $x_1, \ldots, x_n$ of exogenous variables and a noise component ε, in the sense that

$$y = a_1x_1 + a_2x_2 + \cdots + a_nx_n + \varepsilon.$$

In the tide analogy one can make a sharp distinction between two different types of influence: the astronomical components play the role of the x_is and the meteorological components play the role of the noise component, ε. In economics, however, many of the x_is may be difficult to measure, even by way of some proxy variable that we might use to control for their effects. "We are stuck with the fact that some of our systematic influences have slipped into our estimated 'residuals,' that is, into the noise component" (p. 21).

> I believe that much of the difficulty economists have encountered over the past fifty years can be traced to the fact that the economic environments we seek to model are sometimes too messy to be fitted into the mold of a well-behaved, complete model of the standard kind. It is not generally the case that some sharp dividing line separates a set of important systematic influences that we can measure, proxy, or control for, from the many small unsystematic influences that we can bundle into a "noise" term. (Sutton 2000, p. 32)

Although I share Sutton's analysis of the difficulties a field science has to deal with to be an exact science, I disagree with his presupposition that this "standard paradigm" is Haavelmo's. He knew the limitations of such a program too well:

> Certainly we know that decisions to consume, to invest, etc., depend on a great number of factors, many of which cannot be expressed in quantitative terms. What is then the point of trying to associate such behavior with only a limited set of measurable phenomena, which cannot give more than an incomplete picture of the whole "environment" or "atmosphere" in which the economic planning and decisions take place? (Haavelmo 1944, p. 3)

As an "apprentice" of Ragnar Frisch (Bjerkholt 2005), he was too much aware of the limitations of the usage of statistics only, as taught by Frisch:

> I do not claim that the [statistical] technique developed in the present paper will, like a stone of the wise [sic], solve all the problems of testing "significance" with which the economic statistician is confronted. No statistical technique, however, refined, will ever be able to do such a thing. The ultimate test of significance must consist in a network of conclusions and cross checks where theoretical economic considerations, intimate and realistic knowledge of the data and a refined statistical technique concur. (Frisch 1934, p. 129)

Haavelmo's epistemology was much more refined than Sutton suggests. This epistemology is to be found in chapter 2, "The Degree of Permanence of Economic Laws," of his *Probability Approach*, which indeed takes account of the limitations of the tide analogy. Let us assume that for each individual's consumption is explained by the equation

$$y^* = f(x_1, x_2, \ldots, x_n).$$

Haavelmo gave the following clarification of why this explanation of the consumption of each individual will be incomplete:

> We shall find that two individuals, or the same individual in two different time periods, may be confronted with exactly the same set of specified influencing factors x (and, hence, they have the same y^* [by the preceding equation]), and still the two individuals may have different quantities y, neither of which may be equal to y^*. We may try to remove such discrepancies by introducing more "explaining factors," x. But usually, we shall soon exhaust the number of factors which could be considered as *common* to all individuals, and which, at the same time, were not merely of negligible influence upon y. The discrepancies $y - y^*$ for each individual may depend upon a great variety of factors, these factors may be different from one individual to another, and they may vary with time for each individual. (Haavelmo 1944, p. 51)

Two other pioneers of the "standard paradigm," Jan Tinbergen and Tjalling Koopmans, the Dutch compeers of Frisch and Haavelmo, also recognized that statistics is not sufficient by itself to turn economics into an exact science.

4.2. The Modest Role of the Statistician

The awareness that by statistical techniques only—at that time mainly linear regression analysis—one could never arrive at a complete list of causal factors was part of the common understanding of what one could do with statistics among these four men in the late 1930s. Tjalling Koopmans's dissertation, "Linear Regression Analysis of Economic Time Series," is exemplary in this regard. It was issued as a Netherlands Economic Institute (NEI) publication in 1937. In the preface, the directors of NEI justified their decision to publish it by explaining the methodological problem one has to deal with in economics in contrast to a laboratory science:[5]

> One of the great difficulties in the statistical testing of economic laws is the famous "ceteris paribus" clause with which theory has to be formulated. The statistician wishing to isolate a given law has to eliminate changes of

> these other factors assumed to be constant in that law. The most effective implement for this sort of problems is the method usually called multiple correlation analysis, which method has been applied in several investigations in the last few years. It is still an open question, however, whether the results thus obtained have real significance and, if so, to what degree the figures calculated are trustworthy. (Quoted in Koopmans 1937, p. v)

To deal with the impossibility of keeping everything the same while at the same time assessing the "significance" of a regression coefficient, Koopmans suggests the following division of labor between "the economist" and "the statistician." The economist "should by economic reasoning and general economic experience—or by his knowledge of the special branch of science concerned—devise a set of determining variables which he expects to be a complete set" (p. 57).

A set $x_1, x_2, \ldots, x_n$ was defined by Koopmans to be complete if the combined influence on y of all other variables, not included in the set, is represented by a relatively "small summand" in y of "accidental nature" (cf. the noise term in Sutton's example). "If, in any concrete situation, a standard is established for what should be understood by a 'small summand of accidental nature,' the notion of a complete set of determining variables is fixed by that standard" (p. 6).

The statistician, however, cannot test the completeness of this set, because, as Koopmans argued, "the possibility remains that the supposed complete set, though differing from the real one in at least one variable, figures in the regression equation as a representative of it, and is able to do so because of close interrelations, 'by chance' or otherwise conditioned, between variables of the two sets" (p. 57). In order to decide what may occur "by chance," Koopmans saw that the statistician's task requires knowledge of the laws governing the behavior of the variables concerned.

Although the statistician cannot confirm that the economist has correctly identified the complete set of relevant variables or the laws underlying their behavior, for some cases he or she may be able to show that the economist is wrong—namely, for those cases in which a regression equation "expressing the dependent variable in terms of the determining variables of the supposed complete set, fitted to the data, leaves large residuals" (p. 58). In the case of small residuals which do not exhibit systematic variation,

> the task of the statistician is confined to a study of the reliability of the empirical regression coefficients on the hypothesis that the economist indicated the right set of determining variables, or, at least, that he did not omit important determining variables from his list. Such a study is, indeed, possible. (Koopmans 1937, p. 58)

The task of the statistician was modest, in the sense that it was restricted to an evaluation of the "residual." A large residual indicates that the set of causal factors is incomplete, but not what the omitted factor was. A small residual seems to suggest that the set is not incomplete, but that could also be a matter of chance.

In response to the famous Keynes-Tinbergen debate, Koopmans (1941) made the two different and separated tasks of the "economist" and the "statistician" more explicit.[6] The paper was written with the intent to provide "a more systematic exposition of the logic of the methods applied in econometric business-cycle research" (p. 157). This econometric business-cycle research Koopmans referred to was Tinbergen's (1939a, 1939b) research for the League of Nations that led to the two-volume *Statistical Testing*, and which was thoroughly criticized by Keynes (1939).[7] Of interest with respect to one of the general themes of this book—the role of expert judgments—I would like to note here that Koopmans presented the whole discussion of Tinbergen's method again in terms of two human experts, the economist and the statistician, and not in terms of "economic theory" and "statistics," on which more in chapter 6.

The "logic" of econometrics, according to Koopmans, consists of the following "elements." The first element is "the availability of a considerable number of statistical time series, to be referred to as the data, each series representing some measurable economic phenomenon or variable which plays a role in cyclical fluctuations" (p. 158). The second element is

> the adoption, by the investigator, of the working hypothesis that causal connections between the variables are dominant in determining the fluctuations of the internal [endogenous] variables, while, apart from external [exogenous] influences readily recognized but not easily expressible as emanating from certain measurable phenomena, mere chance fluctuations in the internal variables are secondary in quantitative importance. (Koopmans 1941, p. 160)

As examples of "unmeasurable" exogenous factors, Koopmans mentioned earthquakes, political events, and strikes. Koopmans, however, emphasized that this "working hypothesis" should not be understood as excluding immeasurable endogenous factors.

On the basis of these two elements, only negative inferences could be drawn, namely, the inference that a supposed causal connection is not in agreement with the data, which is according to Koopmans an accurate characterization of statistical testing of business cycle theories. But this statistical inference is inconclusive. To make a choice between the numerous possible explanations, one needs "additional information" (p. 162).

It is the "economist's task" to come up with this needed additional information. This "economist" is

> not supposed to be of the too academic type versed only in abstract deduction from the "economic motive." He is considered to have in addition

> an intimate knowledge of economic life and of the results of statistical investigations relating to similar countries and periods. (Koopmans 1941, p. 166n)

The economist's task implies specification of the variables that may (but need not) have an influence on the variable to be explained. Koopmans did not specify what he meant by "intimate knowledge of economic life," but he did say more explicitly what he meant by "additional information": a set of statements that include "observations not expressible as statistical time series, experiences from other countries or periods showing a similar economic structure, deductions from economic theory, or even mere working hypotheses having a certain degree of plausibility" (p. 163). These statements are "not held to be incorrect. Very often, however, they may on good grounds be supposed to approach the truth closely" (p. 166).

The statistician "applies the type of reasoning and the procedures elaborated in statistical theory" (p. 166). The statistician's task is "technical and bears no responsibility for the conclusions other than that it must avoid errors of reasoning" (p. 178). So modeling is "a continuous dialogue, of a game of give and take, between economist and statistician" (p. 178). Due to this "specialization" of tasks, putting together the list of influences to be included in a model is not a task for the statistician but for the economist.

4.3. Significance

Haavelmo, being acquainted with the latest developments in econometrics, was studying what it meant for a "determining variable" to be "important."[8] In this period—the late 1930s and early 1940s—he was clearly investigating this issue, evidenced by three papers written at that time, and which will be discussed here.

The first paper in which he discussed the determination of whether a variable was explanatory significant was a lecture given at the Third Nordic Meeting of Younger Economists, in Copenhagen in May 1939. In this lecture Haavelmo ([1939] 2008) discussed the problem of significance in a very specific way: Even if there is "statistical agreement" between the observed data and the regression equation, it is still uncertain whether the regression coefficients are "significant," that is, whether the absolute values of the coefficients are large relative to their standard errors (see p. 14).

Haavelmo may have received this specific idea of "significance," that is beside "fit" also a relatively large absolute value of the coefficients, when he visited Tinbergen at the League of Nations in Geneva. There is some evidence that Tinbergen, in turn, got the idea from Ragnar Frisch in an exchange of letters discussing Tinbergen's project of econometric modeling at the League of Nations.

In a letter (dated 15 September 1938) to Tinbergen, Frisch discussed what he called "the cyclical intensity coefficient of one variate [variable] with respect to another."[9]

> We speak of one variate as being "more important" than another in explaining the cyclical course of a third variate. What is meant by this? Quantitatively speaking it would have to be expressed by the magnitude of a certain regression coefficient if the data are analysed by means of regressions. But how are we to measure the magnitude of a regression coefficient? (Frisch 1938)

In this letter, Frisch came up with the following suggestion for the determination of a "coefficient whose magnitude has an economic meaning." Suppose we have determined a regression equation of the form $x_0 = b_1x_1 + b_2x_2 + \cdots + b_nx_n$, and one wants to know "how strong" the influence of x_1 on x_0 is. Therefore, Frisch suggested, one puts $x_2 \ldots x_n$ equal to the average values of these variables over the period considered. Then let x_1 change over such a range as it usually passes through during one of those cycles that have been observed.

This change in x_1 corresponds to a change in x_0:

$$x_0^{1max} \equiv b_1x_1^{max} + b_2\bar{x}_2 + \cdots b_n\bar{x}_n$$

$$x_0^{1min} \equiv b_1x_1^{min} + b_2\bar{x}_2 + \cdots b_n\bar{x}_n,$$

where x_1^{max} and x_1^{min} are the highest and lowest values of x_1 in a "normal" cycle. The difference between the two x_0 values may now be compared with the difference between the highest and lowest x_0 values that are actually observed over the period in question. Frisch called the ratio between these two differences "the cyclical intensity of phenomenon no. 0 with respect to no. 1":

$$\frac{x_0^{1max} - x_0^{1min}}{x_0^{max} - x_0^{min}} = b_1 \frac{x_1^{max} - x_1^{min}}{x_0^{max} - x_0^{min}}.$$

It expresses the fraction of the "usual" x_0 amplitude that "would have been produced" if only x_1 had been the "acting cause." To Frisch the "cyclical intensity" seemed a "very plausible measure."

> At any rate the measure will yield it seems an interesting classification between "important" and "less important" factors. Personally I have felt it difficult to interpret how strong the various effects expressed by your equations are. It would be a great help if for each equation (or rather

> for each term of each equation) the corresponding intensity coefficient—according to some such definition as the above—could be given. (Frisch 1938)

I am not aware of any direct response from Tinbergen to Frisch's letter, so it is not clear how it influenced any of the revisions of the volumes of *Statistical Testing of Business-Cycle Theories*. One can find, however, a definition of the "strength of influence" that does not appear in any of Tinbergen's earlier publications.

Tinbergen defined the term b_2x_2 in an expression like $x_1 = b_2x_2 + b_3x_3 + \cdots$, as the "influence of x_2." The "strength of the influence of x_2" was defined as $b_2\sigma_2$, that is, the standard deviation of x_2 times the regression coefficient (Tinbergen 1939a, pp. 22–23). And subsequently, Tinbergen used this idea of strength in one of his many tests to evaluate the most important explanatory variables of "investment activity."[10]

This evaluation of the explanatory variables based on the regression analysis was presented in a section titled "Details of Results, First Stage." In three tables (covering the United Kingdom, Germany, and the United States), labeled as "'Explanation' of Investment Activity, First Stage," the correlation and regression coefficients of the main variables were presented. The selection of the variables was based on the following coefficients: on their correlation coefficient with investment activity and on their "influence." Variables showing "wrong signs" or "a very small influence" were to be excluded. An "influence" of a variable on a dependent variable was measured by multiplying the regression coefficient of these two variables by the standard deviation of that particular influence (see Tinbergen 1939a, pp. 51–53).

As Tinbergen explained, the tables were given for three purposes: (1) the analysis of the correlation coefficient of a new variable and the variable to be explained, (2) the analysis of the stability of the regression coefficient in various cases, and (3) the analysis of the "relevance" of the variables. This latter analysis was based on

> i. whether or not the variate [variable] in question increases the correlation coefficient to any considerable extent,
> ii. whether or not the sign of a fairly stable regression coefficient is right, and
> iii. whether or not the influence of that variate is perceptible. (Tinbergen 1939a, p. 50)

Tinbergen did not explicitly say what he meant by "perceptible," but what he probably meant was that the influence is large enough, without providing a metric for its largeness. It was indeed judged by perception only.

As a result of this analysis, Tinbergen concluded that "the importance of profits (Z), and, in the case of the United States, of share yield (m_{LS}), is confirmed: the increase in correlation and the influence of the variate are considerable; the signs are right" (p. 54). In other words, in the analysis to arrive at this conclusion, besides "correlation" and "sign," "influence" played an equal role in deciding the "relevance" of a variable in a model.

In the second volume, which was produced after the first, Tinbergen (1939b, p. 13) was, however, more hesitant to use the "strength of influence" (regression coefficient times standard deviation) as a test for the significance of a variable:

> It goes without saying that if some explanatory factor has not changed at all in the period studied, its influence cannot be determined. If it changed only slightly, its regression coefficient may be uncertain. Extrapolation of such results for large variations in the factors concerned is therefore not permitted.

It may be that Haavelmo had a role in this later move to a more hesitant use of the "strength of influence."

That Haavelmo probably had doubts about the use of the "strength of influence" as a measure for significance and that he probably had revealed these doubts to Tinbergen can be inferred from two of Haavelmo's publications at that time. In his 1939 Copenhagen lecture, to discuss whether the "strength of influence" could play a role in evaluating whether a causal factor was relevant, Haavelmo made a distinction between two types of errors, each quantifying in a specific way the error of omitting a variable in an explanation. They were called the "error of the momentary explanation" and the "error of the average relation."[11]

He showed the difference between these two types of error with the following example: Consider N observations on three variables, x, y and z, between which exists the following true but unknown relation (note that the three variables are measured from their respective means):

$$y = 2x - 3z\,. \tag{4.1}$$

Suppose we assume that a relationship exists between y and x but we are ignorant about the influence of z, so z is omitted:

$$y = bx\,.$$

The correlation coefficient between y and x is

$$r_{xy} = 2\frac{\sigma_x}{\sigma_y} - 3r_{xz}\frac{\sigma_z}{\sigma_y}\,.$$

As a result the regression coefficient b is

$$b = 2 - 3r_{xz}\frac{\sigma_z}{\sigma_x}\,.$$

Haavelmo defined the "error of the average relation" as the standard error of the regression coefficient b:

$$\sigma_b = \frac{1}{\sqrt{N-2}} \frac{\sigma_y}{\sigma_x} \sqrt{1 - r_{xy}^2}\,. \tag{4.2}$$

This error tells us that, despite the omission of z, the larger the number of observations (N), the more precisely the regression between y and x is determined. So, according to Haavelmo, the larger the N, the "more stable" the "average relation" between y and x. But this does not mean that this relation is "a good description of the momentary variation in y" (p. 15).

He defined the "error of the momentary relation" as the mean squared difference between the observed y and the estimated, called the "calculated" y, denoted by $\hat{y}$:

$$\frac{1}{N}\sum_{i=1}^{N}(y_i - \hat{y}_i)^2 = \frac{1}{N}\sum_{i=1}^{N}\left(2x_i - 3z_i - 2x_i + 3r_{xz}\frac{\sigma_z}{\sigma_x}x_i\right)^2 = 9(1 - r_{xz}^2)\sigma_z^2.$$

So when the number of observations increases, the error of omitting variable z in the "average explanation" σ_b decreases (see eq. 4.2), but the error of omitting z in the "momentary explanation" remains at the same level of $9(1 - r_{xz}^2)\sigma_z^2$.

This "error in the momentary explanation" was subsequently used to discuss the following cases: The smaller the variation in z, σ_z, and the larger the correlation r_{xz}, the better the "momentary explanation of y." A high correlation r_{xz} means that the variation in x also accounts for a considerable part of the variation in z, so the omission of z in the explanation of y is not that bad. The more interesting case, however, is the opposite case when both the correlation r_{xz} and the variation of z, σ_z, are small:

> This means that z is a superfluous variable when it comes to explaining the observed variation of y in this material [data set]. But that does not necessarily mean that [the regression equation] will give a good forecast of y outside the period covered by the data. (Haavelmo [1939] 2008, p.15)

Thus, for a specific data set it may be the case that z did not show much variation and z and x are not strongly correlated. For this data set z will not show itself sufficiently, and hence will not be included in the explanation. But it would be a mistake to assume therefore that z is not a causal factor, because it is; see equation 4.1.

Equation 4.1 shows that if z varies strongly enough, it would exert a substantial influence. Thus, although x alone may give a good explanation of y for the used data set, "it may be of decisive importance whether or not we can utilize the tiny part of the variation remaining in attempting to capture the effect of z" (pp. 15–16).

> This indeed shows how crucial it is to have, in advance, a formulation of the hypotheses, in which one operates with specific fictitious variations. If one refrains from doing so, one runs the risk of missing out [sic] important variables that, by change or for specific reasons, have not shown significant

> variation in the material at hand. And although a simpler hypothesis may give a stable average explanation and for that reason gets accepted as statistically valid, it may well give a very poor, maybe a completely worthless momentary explanation and no deeper insight into structural relationships between the studied variables. (Haavelmo [1939] 2008, p. 16)

In other words, the "strength of the influence," that is, the regression coefficient times the standard deviation of its variable, is not a good indicator of whether a variable is a deeper, structural factor, because it depends too much on whether the variable has shown significant variation within the period of study, or for a specific data set.

This same issue was also discussed in a more direct criticism of Tinbergen's method employed in his work at the League of Nations. In a 1941 "note," Haavelmo criticizes Tinbergen's method of using regression analysis to select the main causal factors, in particular Tinbergen's decision that the rate of interest does not causally affect investment. Moreover, this contradicted the "classical dynamic theories of production and prices," according to which the interest rate is a "powerful autonomous parameter" (Haavelmo 1941b, p. 49). By calling a factor an "autonomous" parameter, influence, change, or regulator (see below) Haavelmo meant to say that it was a causal factor.

Tinbergen's 1939a article concluded that the rate of interest is "far less important than profits and share yields" (p. 55) because the regression analysis showed that the regression coefficient of the interest rate was not significantly different from zero, and so also its strength of influence (as discussed above). Haavelmo's (1941b) "note," however, aimed to show that the coefficient's being zero "by itself is *not* sufficient to establish this conclusion" (p. 49). Moreover, it aimed to show that regression equations "may be very misleading if the regression coefficients [are] taken to represent *effects of autonomous changes* in the corresponding variables" (p. 49).

Haavelmo showed for a model representing the structural relations with respect to investment activity that even if the interest rate is a "causal component," its regression coefficient can still be zero: Let $x_1, x_2, \ldots, x_n$ be a number of causal components, where x_1 is gross profit, x_2 is total interest expenses, and so on. Let $Y, y_1, y_2, \ldots, y_n$ be the observable variables, where Y is the volume of investment, y_1 is net profit, y_2 is the interest rate, and so on. The regression equation that should fit the data is

$$Y = a_1 y_1 + a_2 y_2 + \cdots + \varepsilon. \tag{4.3}$$

But if

$$Y = k(x_1 + x_2 + \cdots + x_n) + \varepsilon$$

$$y_1 = k_1(x_1 + x_2 + \cdots + x_n)$$

$$y_2 = k_2 x_2$$

then the regression coefficient a_2 is zero. The reason is that net profit, y_1, already includes causal component x_2. But also when $y_2 = k_3x_3$, where x_3 is "some *other* effects of the interest on investment" (p. 50), like the substitution effect, different preference schedules for different kinds of holdings, and so on, a_2 would still be zero, because these effects are partly taken care of by net profit, y_1. This result (that a_2 would still be zero), however, as Haavelmo emphasized, does not imply that changes in the interest rate have no influence upon investment; the interest rate "may *indirectly* exercise a decisive influence" (p. 50) upon Y, namely via its influence upon y_1. With this example Haavelmo showed that a causal structure can be such that regression analysis does not show what the causal factors are.

Besides the fact that the causal structure can be such that the regression coefficient of a causal component will always be zero, Haavelmo gave a second reason for the coefficient being zero, which was the same as in his 1939 lecture, but now presented in a much more compact way:

> The rate of interest may not have varied much during the statistical testing period, and for this reason the rate of interest would not "explain" very much of the variation in net profit (and thereby the variation in investment) which has actually taken place during this period. But one cannot conclude that the rate of interest would be inefficient as an autonomous regulator, which is, after all, the important point. (Haavelmo 1941b, p. 50).

This relation between regression analysis and how we know whether a factor is a causal factor will be discussed in the next section.

To find the "efficiency" of the "autonomous regulator," Frisch's (1938) suggested method of determining the "intensity coefficient" does not help, either. This method was based on the application of ceteris paribus assumptions to determine the "strength of an influence" (as discussed previously).

> In economic theory we discuss the effect of interest-rate changes upon investment under the assumption of "other things being equal." Statistically, we may try to fulfil this *ceteris paribus* requirement by including the most important of these "other things" in a regression equation, in order to see the effect of interest-rate changes when these "other things" are *kept constant*. But the statistically defined "other things" must correspond to those we have in mind in theory. The condition [y_1 = constant in equation 4.3] is evidently *not* the *ceteris paribus* condition we have in mind in theory; in fact, it is the very opposite, because [y_1] = constant means such a particular change in all the other factors entering into net profit that the effect of an interest change is exactly *cancelled out*. (Haavelmo 1941b, p. 50)

Assume we have found the following regression equation:

$$Y = a_1 x_1 + a_2 x_2 + \cdots + a_n x_n,$$

and we would like to determine the strength of the influence of the interest rate, x_2, on investment, Y, by imposing ceteris paribus conditions on the other variables in this equation (where Δ denotes change or variation):

$$\Delta Y = a_2 \Delta x_2 \quad \Delta x_1 = \Delta x_3 = \cdots = \Delta x_n = 0.$$

Then, however, we would obtain a spurious result because the ceteris paribus condition would conflict with theory: To keep x_1 constant, $y_1, y_2, \ldots, y_n$ should vary in such a way that they cancel each other out. But then we have $\Delta Y = 0$!

4.4. Haavelmo's Epistemology

The epistemological insights of the 1939 Copenhagen lecture and his note on the limitations of regression analysis with respect to the recovery of causal factors were generalized in Haavelmo's 1941 treatise "On the Theory and Measurement of Economic Relations," which was published in 1944 as the *Probability Approach in Econometrics*. In particular, the section titled "The Question of Simplicity in the Formulation of Economic Laws" discusses the problem of arriving at a complete explanation of certain economic behavior or phenomena.[12] This more general framework allowed him to discuss issues about finding laws outside ceteris paribus environments.

Haavelmo did not share Mill's and Cartwright's pessimism; as he noted: "a phrase such as 'In economic life there are no constant laws,' is not only too pessimistic, it also seems meaningless" (p. 14). Instead, Haavelmo suggested, "We may try to find a rational explanation for the fact that relatively few attempts to establish economic 'laws' have been successful" (p. 14).

The "rational explanation" he developed in his dissertation can be clarified by first considering the following generalization of his analysis presented in the Copenhagen lecture: Note again that the variables are measured from their respective means. Assume that the true relation between y and a set of independent causal factors (correlation coefficients between them are zero) is

$$y = a_1 x_1 + a_2 x_2 + \cdots + a_n x_n.$$

Assume that one of the variables, say x_i, does not deviate from its mean ($\sigma_i = 0$), that is, it remains constant for the considered data set. Then the correlation of variable x_i, $r_{iy} = a_i \sigma_i / \sigma_y = 0$ and thus the regression equation is

$$\hat{y} = a_1 x_1 + \cdots + a_{i-1} x_{i-1} + a_{i+1} x_{i+1} + \cdots + a_n x_n.$$

As a result, the "error of the momentary explanation" is

$$\frac{1}{N}\sum_{k=1}^{N}(y_k - \hat{y}_k)^2 = a_i^2\sigma_i^2.$$

In his later 1941/1944 assessment of this result, Haavelmo saw that this "error" could be reinterpreted as a measure of the "factual influence" of the variable x_i on y, which he defined as follows: Let y be a theoretical variable defined as a function of n independent causal variables $x_1, x_2, \ldots, x_n$:

$$y = f(x_1, x_2, \ldots, x_n).$$

Let us replace the variable x_i by a constant c_i so determined that

$$Q_i = \sum_{j=1}^{N}\left[f\left(x_{1j},\ x_{2j},\ \ldots,\ x_{ij},\ \ldots,\ x_{nj}\right) - f\left(x_{1j},\ x_{2j},\ \ldots,\ c_i,\ \ldots,\ x_{nj}\right)\right]^2$$

is a minimum with respect to c_i. The factual influence of the variable x_i upon y is then defined as Constant$\cdot\sqrt{Q_i^{(min)}}$ (p. 24).

To see that the factual influence is indeed a generalization of the error in the momentary explanation, take the following linear case: $f(x_1, x_2, \ldots, x_n) = a_1x_1 + a_2x_2 + \cdots + a_nx_n$. Then $c_i = \bar{x}_i$ and thus $\frac{1}{N}\sqrt{Q_i^{(\min)}} = a_i^2\sigma_i^2$.

To arrive at a more general framework to discuss the relevance of causal factors, Haavelmo defined the "potential influence" of a causal factor x_i upon y as (p. 23)

$$\Delta_i y = f(x_1, x_2, \ldots, x_i + \Delta x_i, \ldots, x_n) - f(x_1, x_2, \ldots, x_n).$$

To compare the size of the potential influence of each of the variables x_i, one has, for any point $(x_1, x_2, \ldots, x_n)$, to choose a set of displacements $\Delta x_1, \Delta x_2, \ldots, \Delta x_n$, which are considered to be of "*equal size* according to some *standard of judgment*" (p. 23).

This concept of potential influence can be clarified by using the definition of a partial derivative:

$$\frac{\partial f}{\partial x_i} \approx \frac{f(x_1, x_2, \ldots, x_i + \Delta x_i, \ldots, x_n) - f(x_1, x_2, \ldots, x_i, \ldots, x_n)}{\Delta x_i}.$$

For a fixed set of displacements, say $\Delta x_i = \Delta_i$, the potential influence can be rewritten as

$$\Delta_i y = \frac{\partial f}{\partial x_i}\Delta_i.$$

As we can infer from this expression, the potential influence is a feature of the function f, or as Haavelmo put it, "For a given system of displacements $\Delta x_1, \Delta x_2, \ldots, \Delta x_n$, the potential influences are clearly, formal properties of the function f" (pp. 23–24).

In the Copenhagen lecture Haavelmo made a few remarks that confirm my interpretation of potential influence in terms of partial derivatives. In a section on the ceteris paribus clause, he distinguishes between two types of ceteris paribus clauses. The first is the usual type where in the selection of essential "elements" guided by theory and data, one imposes the ceteris paribus clause on all remaining unspecified elements "*because* the total effect of these other elements has *by experience* played no large part for the problem at hand and neither can they be expected to do so in the future, so that whether we assume them unchanged or let them play freely is virtually of no consequence" (Haavelmo [1939] 2008, p. 9). This ceteris paribus clause is actually a *ceteris neglectis* clause, saying that the other influences are negligible. This clause will be discussed in more detail below.

Of more interest for the interpretation of "potential influence" in terms of partial derivatives is the second type of ceteris paribus he discussed,

> the one we impose *within* the system of specified variables. The idea here is just the same as that of partial derivatives. We study relations between some among the specified objects, which are mutually independent, subject to the assumption that the remaining elements specified are kept constant. Usually, the form that such relations takes depends on the *level* at which the other elements are fixed. Such reasoning is not only of theoretical interest, on the contrary, it is also the basis for assessing effects of practical measures of intervention in the economic activity. (Haavelmo [1939] 2008, p. 9)

This latter aspect of assessing the effects of interventions by using partial derivatives will be discussed now.

Using the expression for potential influence, we can now rewrite the expression for the factual influence as[13]

$$\frac{\partial f}{\partial x_i}\Delta x_i.$$

In other words, the factual influence consists of two components: potential influence and variation. What we (passively) observe is the factual influence of a causal factor, which is only the case when that factor has sufficiently large potential influence ($\partial f/\partial x_i >> 0$) and has varied sufficiently ($\Delta x_i >> 0$).[14] Therefore, potential influence of a causal factor can only be measured indirectly, that is, if that factor has shown factual influence. In other words, the magnitude of the first component is only known if the second component (that is, the variation) is significantly different from zero.

With these two concepts of potential influence and factual influence, Haavelmo could now better clarify the limits of statistical inferences. In economics, the investigator is a "passive observer," which means that "he is not in a position to enforce the prescriptions of his own designs of ideal experiments" (p. 7). It is not only the impossibility of enforcing an experimental design; "passive observation" also refers to the impossibility of intervention or control. In a paper "The Problem of Testing Economic Theories by Means of Passive Observations," which he presented at the Cowles Commission's 1940 Research Conference, he emphasized this latter characteristic:

> Can we measure economic structure relations (e.g., individual indifference surfaces or other "behavioristic" relations) by means of data which satisfy simultaneously a whole network of such relations, i.e., data obtained by a "passive watching of the game" and not by planned experiments? (Haavelmo, 1940, p. 59)

The problem then is how to find laws by passive observations only:

> How far do the hypothetical "laws" of economic theory in its present stage apply to such data as we get by passive observations? By passive observations we mean observable results of what individuals, firms, etc., *actually do* in the course of events, not what they *might* do, or what they *think* they would do under certain other specified circumstances. (Haavelmo 1944, p. 16)

Laws refer to hypothetical situations of what might happen, whereas observations are related that what actually happens. Or in other words, laws are "describing a schedule of alternatives at a given moment, *before* any particular decision [on what to do] has been taken," whereas the relations we find by regression analysis are "those intended to describe what the individuals *actually do* at any time" (p. 18).

To acquire knowledge about laws we need to acquire knowledge about the potential influences that are accounted for in a law. Each potential influence significantly different from zero is a relevant causal factor. To gain this knowledge, Haavelmo distinguishes two different classes of experiments:

> (1)experiments that *we should like to make* to see if certain real economic phenomena—when *artificially isolated* from "other influences"—would verify certain hypotheses, and (2) the stream of experiments that Nature is steadily turning out from her own enormous laboratory, and which we merely watch as passive observers. (Haavelmo 1944, p. 14)

In case of the first class of experiments, to isolate a selected set of factors, $x_1, \ldots, x_n$, from the other influences, $x_{n+1}, x_{n+2}, \ldots$, we first impose ceteris paribus conditions on the latter set—$\Delta x_{n+1} = \Delta x_{n+2} = \cdots = 0$—so that a simpler relationship can be investigated:

$$\Delta y_{CP} = \frac{\partial f}{\partial x_1}\Delta x_1 + \cdots + \frac{\partial f}{\partial x_n}\Delta x_n.$$

In such a controlled experiment, the remaining factors, x_i, can be varied in a systematic way to gain knowledge about the $\partial f/\partial x_i$'s and so to see whether they are stable for these variations.[15] If this applies for all factors $x_1, \ldots, x_n$, we have found a lawful relation between y and these factors.

In the second class of experiments, those conducted by Nature, we usually passively observe a limited number n of factors that have a nonnegligible factual influence:

$$\Delta y_{PO} \approx \frac{\partial f}{\partial x_1}\Delta x_1 + \cdots + \frac{\partial f}{\partial x_n}\Delta x_n.$$

(the other factors have a negligible influence: $\frac{\partial f}{\partial x_{n+1}}\Delta x_{n+1} \approx \frac{\partial f}{\partial x_{n+2}}\Delta x_{n+2} \approx \cdots \approx 0$). Thus, the relationship $y = f(x_1, \ldots, x_n)$ explains the actual observed values of y, provided that the other factual influences of all the unspecified factors are negligible.

The problem with passive observations, however, is that it is not possible to identify the reason for the factual influence of a factor, say x_{n+1}, being negligible, $(\partial f/\partial x_{n+1})\Delta x_{n+1} \approx 0$. As passive observers, we cannot distinguish whether its potential influence is very small, $\partial f/\partial x_{n+1} \approx 0$, or whether the factual variation of this factor in the data set period under consideration was too small, $\Delta x_{n+1} \approx 0$. We would like only to omit the factors whose influence was not observed because their potential influence was negligible to start with. At the same time, we want to retain factors whose influence was not observed because they varied so much less that their potential influence was veiled, but may do so in the future, which would invalidate our explanation (p. 26).

To know whether a certain factor is a potential influence, and should therefore be taken account of, we are dependent on the kind of experiments Nature has conducted so far. Regression analysis is based on these variations shown by Nature in the past and therefore does not give knowledge about whether a set of theoretically suggested causal factors is complete. There may always be a factor that will only start to vary in the future.

This problem of passive observation can be tackled in two ways. One is the accumulation of data sets with the expectation that an increasing number of potential influences become visible; the other is to take account of as many as causal factors as theories suggest.

> Frequently, our greatest difficulty in economic research does *not* lie in establishing simple relations between actual observation series, but rather in the fact that the observable relations, over certain time intervals, *appear to be still simpler* than we expect them to be from theory, so that we are thereby led to *throw away* elements of a theory that would be sufficient to explain apparent "breaks in structure" later. (Haavelmo 1944, p. 26)

How can this framework help us to "establish constant laws of economic life" (p. 17), where one has to deal with "ever-shifting environments" (p. 17)? Haavelmo's framework shows that besides ceteris paribus environments, $\Delta x_{n+1} = \Delta x_{n+2} = \cdots = 0.$, laws can also exist in ceteris neglectis environments. These are environments in which all kinds of background influences are constantly changing, but the "permanence" of a law is not affected because the potential background influences are all negligible, $\partial f/\partial x_{n+1} \approx \partial f/\partial x_{n+2} \approx \cdots \approx 0$.

4.5. Theory or Other Additional Information?

Like Sutton, Haavelmo did not believe that economics could be turned into an exact science.

> From experience we know that attempts to establish *exact* functional relationships between observable economic variables would be futile. It would indeed be strange if it were otherwise, since economists would then find themselves in a more favorable position than any other research workers, including the astronomers. Actual observations, in whatever field we consider, will deviate more or less from any exact functional relationship we might try to establish. (Haavelmo 1944, p. 40)

But although he did not believe that economics could be an exact science, Haavelmo did not conclude that we should give up our "hope to find elements of invariance in economic life, upon which to establish permanent 'laws'" (p. 13). Haavelmo's framework "may give us some hint as to how optimistic or pessimistic we have reason to be: we can try to indicate what would have to be the actual situation in order that there should be *no hope* of establishing simple and stable causal relations" (p. 23). It shows that for the existence of laws we do not need to have the hard requirement of ceteris paribus environments, $\Delta x_{n+1} = \Delta x_{n+2} = \cdots = 0$, but we can do with the softer requirement of a *ceteris neglectis* environment, $\frac{\partial f}{\partial x_{n+1}} \approx \frac{\partial f}{\partial x_{n+2}} \approx \cdots \approx 0$.[16]

This does not deny the other part of the problem of finding laws outside the laboratory, namely, how to gain knowledge about potential influences, to know which ones are relevant and stable. If Nature does not tell us much, that is, by varying too few of them, theories may help—"it is a task of making fruitful hypotheses as to how reality actually is" (p. 31)—and the accumulation of data sets may also help. The same problem was also discussed by Sutton, but without referring to Haavelmo's work:

> We need to have a data set in which the values of x_i fluctuate widely, thus leaving a clear trace of x_i's influence on y. In practice, we will work with a limited set of data, and many of the potentially relevant factors may show

> little variability. The estimated form of the equation may indicate that only a limited subset of the x_i's are significantly different from zero. (Sutton 2000, p. 19)

And if theory does not help us either, experience, may be also a helpful source:

> It is a creative process, an art, operating with rationalized notions of some real phenomena and of the mechanism by which they are produced. The whole idea of such models rests upon a belief, already backed by a vast amount of experience in many fields, in the existence of certain elements of invariance in a relation between real phenomena, provided we succeed in bringing together the right ones. (Haavelmo 1944, p. 10)

In the Cowles Commission approach to econometrics developed later, based on Haalvelmo's *Probability Approach*, theory came to have the prominent and sole task of providing a complete list of potential influences; "little attention was given to how to choose the variables and the form of the equations; it was thought that economic theory would provide this information in each case" (Christ 1994, p. 33). Also in the work of Koopmans the "economist" finally came to be replaced by "theory," as one, for example, can see in a paper jointly written with Herman Rubin and Roy B. Leipnik, "Measuring the Equation System of Dynamic Economics":

> The analysis and explanation of economic fluctuations has been greatly advanced by the study of systems of equations connecting economic variables. The construction of such a system is the task in which economic theory and statistical method combine. Broadly speaking, considerations both of economic theory and of statistical availability determined the choice of the variables. (Koopmans, Rubin, and Leipnik 1950, p. 54)

But more explicitly Koopmans emphasized the crucial role of theory in measurement, in his famous "Measurement without Theory" article in 1947, which was a review of Arthur F. Burns and Wesley C. Mitchell's book *Measuring Business Cycles* (1946). He accused Burns and Mitchell of trying to measure economic cycles in the absence of any theory about the workings of such cycles: "The toolkit of the theoretical economist is deliberately spurned" (Koopmans 1947, p. 163).

His main argument to explain the implications and limitations of such an "empiricist position" is that for purposes of systematic and large-scale observation of a many-sided phenomenon such as the business cycle, "theoretical preconceptions about its nature cannot be dispensed with, and the authors do so only to the detriment of the analysis" (p. 163). He compared this empiricist position with Kepler's discovery of the more superficial empirical regularities of planetary motion, which fell short of the deeper "fundamental laws" later discovered by Newton.

Newton's achievement was based not only on the regularities observed by Kepler, but also on experiments conducted by Galileo. However, Koopmans believed that economists are unable to perform experiments on an economic system as a whole. According to Koopmans,

> It is therefore not possible in many economic problems to separate "causes" and "effects" by varying causes one at a time, studying the separate effect of each cause—a method so fruitful in the natural sciences. On the other hand, economists do possess more elaborate and better established theories of economic behavior than the theories of motion of material bodies known to Kepler. (Koopmans 1947, p. 166).

Thus, to acquire causal knowledge economists have the availability of theory instead of experiment, which does not make them worse off than physicists; quite the contrary, economists can proceed because the evidence for this is based on "introspection."

The theory Koopmans had in mind was general equilibrium theory, but the relevant question is whether this theory is rich (that is, "complete") enough to account for all potential influences. Not according to Rutledge Vining, who replied, on behalf of the NBER, to Koopmans's accusations: "Is the Walrasian conception not in fact a pretty skinny fellow of untested capacity upon which to load the burden of a general theory accounting for the events in space and time which take place within the spatial boundary of an economic system?" (Vining 1949, p. 82).

According to Vining, Haavelmo's *Probability Approach* gives a fourfold classification of the main problems in quantitative research: "first, the construction of tentative theoretical models; second, the testing of theories; third, the problem of estimation; and fourth, the problem of prediction" (p. 82). Hereby he noted that the first problem is "the only one of the four that is not a problem of strictly statistical theory" (p. 82). Vining asserted that the theory about the behavior of economic agents had not been given in sufficient detail.

> When we think of the enormous body of factual knowledge digested and systematized in other fields of variation and the meagerness of our own fields of variation from efforts to systematize, are we quite ready to leave Haavelmo's first problem and launch into the last three problems in estimation theory? (Vining 1949, pp. 82–83)

How much theory do we need for measurement? Do we need a theory that provides the complete list of potential influences, or can we do with a more "skinny" theory but one aided with additional information, because economics is—due to the nature of the phenomena it studies—an inexact science? According to Kevin Hoover (1994) we should answer this latter question affirmatively. He therefore sees econometrics as a specific kind of "observation" instead of the kind of measurement that Koopmans was assuming in his attack on the NBER approach, namely as a "direct measurement of structures suggested by economic theories to be replicas of economic reality" (p. 73). Koopmans's kind of measurement presupposes "strong apriorism," where the theory is assumed to be exact or complete. "Observation" as a specific kind of measurement only requires "weak apriorism": theory guides observations but observation can suggest which elements of a theory are unsatisfactory. "Measurement requires prior theory; equally, theory requires prior measurement" (p. 73), which is according to Hoover the position expressed by Haavelmo.[17] The problem of finding a complete list of potential influences is

> a problem of actually knowing something about real phenomena, and of making realistic assumptions about them. In trying to establish relations with high degree of autonomy we take into consideration various changes in the economic structure which might upset our relations, we try to dig down to such relationships as actually might be expected to have a great degree of invariance with respect to certain changes in structure that are "reasonable." (Haavelmo 1944, p. 29)

Hoover (2002a) explains Cartwright's pessimism about finding economic laws by seeing in her view Koopmans's strong apriorism when it comes to economic measurement. If so, that would be an "irony" in her view, because for physics she does not require this. In her *How the Laws of Physics Lie* (1983, p. 13), she claims that the fundamental laws are not exact. Otherwise,

> they should give a correct account of what happens when they are applied in specific circumstances. But they do not. If we follow out their consequences, we generally find that the fundamental laws go wrong; they are put right by the *judicious corrections* of the applied physicist or the research engineer. (Emphasis added)

In her *Nature's Capacities and Their Measurement* (1989, p. 8), Cartwright reiterates the same claim:

> It is hard to find [laws] in nature and we are always having to make excuses for them: why they have exceptions—big or little; why they only work for models in the head; why it takes an engineer with a special knowledge of real materials and a not too literal mind to apply physics to reality.

While Cartwright acknowledges that practical physics requires knowledge of local regularities, often without a deep or consistent theoretical base, Hoover (2002a, p. 52) argues that "she fails to note the importance of local regularities in economics or to entertain the thought that they might be established econometrically." He agrees with Cartwright that universal regularities are genuinely hard to find because they are rare, but local regularities are ubiquitous: "It would be difficult to make our way through the world if that were not true" (p. 52). And knowledge about these regularities can be used to complete our models.

Hoover, as Haavelmo, thus ends up with a more optimistic view than Sutton and Cartwright about the prospects for econometrics. Econometrics should be seen as an observational science:

> Econometrics is not about measuring covering laws. It is about observing unobvious regularities. . . . Nor can I agree with the message implicit in Cartwright's work that the conditions under which econometrics could succeed are too demanding to be met. The goal of econometrics is not to serve as a nomological machine [highly structured arrangements that generate regularities] nor as its blueprint, but to discover facts that are generated by unobservable nomological machines, facts that theoretical models explain by providing rough drawings, if not blueprints. . . . The robustness of econometric facts is an argument for the existence of nomological machines, but the tools for discovering those facts do not presuppose (fully articulated) knowledge of the construction of those machines. (Hoover 2002b, pp. 173–174)

I agree with Hoover's view that econometrics is closer to Haavelmo's design than Koopmans's. As we saw in chapter 2, current measurement theory does not require strong apriorism, but rather that we acknowledge Type B uncertainty. This still leaves the problem open where this knowledge of local regularities comes from. Although both Haavelmo and Koopmans in his earlier work refer to this kind of knowledge, they never specified what kind of knowledge it should be. This will be discussed in the next chapter.

4.6. Conclusions

In the preface of both the 1941 and 1944 versions of *Probability Approach*, Haavelmo thanks among others Edwin B. Wilson "for reading parts of the original manuscript, and for criticisms which have been utilized in the present formulations" (Haavelmo 1944, p. vi). His use of Wilson's criticism seems, however, not to have gone far enough. In a review of this work, Wilson remains critical, particular about the first two chapters:

> The work is difficult reading ... the author's approach is extremely abstract and metaphysical; [the first two chapters seem] to be written

> quite as much from the point of view of the philosophy of scientific method in general as from that of economic analysis in particular. (Wilson 1946, p. 173)

Ignoring the negative tone of the review, I fully agree with Wilson. Haavelmo's first two chapters provide a very rich epistemological framework for understanding what is entailed in doing scientific research outside the laboratory. Haavelmo's epistemology shows that a social field science explanation can never be made complete by statistical analysis only. For completeness we need "additional information," based on "a vast amount of experience."

Haavelmo's rich epistemology of potential influences became invisible because the idiosyncratic terminology of Frisch and Haavelmo, by which this epistemology was expressed, was replaced by a language whose vocabulary could no longer capture their rich epistemology.

One reason for this change of language is that this epistemology was developed in a period of transition in which early econometrics as data-driven research turned into a theory-guided discipline and in which econometrics became a probabilistic approach. Econometrics in the 1930s as carried out by one of the leading figures of those days, Jan Tinbergen, was an inductive science. In the 1940s, however, econometrics had turned into a deductive science.

An important catalyst of this transition was Haavelmo's methodological manifesto *The Probability Approach in Econometrics* (1944). Although Haavelmo initially was enthusiastic about Tinbergen's project at the League of Nations, he became pessimistic about the achievements of the inductive approach of econometric modeling. The reason for this is that Haavelmo came to understand that observations of what actually happens in an economic system (presented in time series) do not give us sufficient knowledge about the underlying causal mechanism.

The key problem Haavelmo was dealing with is the problem that data observed passively might not display enough variations to reveal the relevant causal influences or relationships of the mechanisms that lie behind the phenomena. In other words, the central problem is that of detecting and measuring the important influences (also called causal factors) of an economic system under investigation when the observations are passive, that is, without the possibility of intervention or control. For this reason, this problem is called the "problem of passive observation." This problem is closely connected to a long list of other tough problems econometricians of the 1930s and 1940s were struggling with: the problem of autonomy, the problem of causality, and the problem of estimation, as well as several others, such as the problem of identification, the problem of invariance, the problem of multicollinearity, the problem of prediction, and the problem of simplicity. Besides these problems are considerably interconnected, another issue is that, during this transition period, the terms used to label them did not have stable meanings. To understand the development of econometrics in the 1930s and 1940s, one must be fully aware that the meanings of the labels often shifted during the period; see for example John Aldrich's historical reconstructions of the development of the term (and theory of) "identification" (Aldrich 1994) and the development of the concept of "autonomy" (Aldrich 1989).

The interconnectedness and changing meanings of these labels may be why the problem of passive observation is underexamined. However, focusing on the problem of passive observation may provide an(other) understanding of the reasons underlying the shift of econometrics from an inductive to a deductive science.

It should be noted that Haavelmo used the term "passive observation" in connection with another problem that could be labeled the problem of simultaneity (see the quotation above from his 1940 paper "The Problem of Testing Economic Theories by Means of Passive Observations"). In the successor literature, this problem came to be known as the problem of identifying structural relationships when all we have are passive observations, and thus it is closely related to the problem of identification.[18] This may be why the discussion about detecting potential influences has vanished in the literature. Nevertheless, I would like to discuss this problem separately from the issue of identification because as a separate problem it clarifies the need for theoretical guidance outside the laboratory when aiming at the recovery of (causal) structure.

Notwithstanding that the language of potential influences has disappeared, it does not mean that Haavelmo's insights have become less important; they merely appeared and still appear in different guises. For example, to identify the correct causal structure it is relevant to know which of the variables have potential influence and which do not. In later discussions of this problem in which the term "potential influence" does not appear, the problem is rephrased in terms of whether certain parameters, for example, regression coefficients, in a particular econometric model are zero and for what reason. Are they zero because for the data set that has been used to estimate them, they appear to be zero, or are they "ontologically" zero, and so indicate that the corresponding variable is not a relevant causal factor? The need for a distinction between the consequences of a factual influence being zero and those of a potential influence being zero is more recently noted by Hoover (2001, p. 45):

> The need for a distinction between a parameter α that is an element of a nonempty set with a range of values, but which happens to equal zero in a particular case, and a parameter α that is an element of the null set ($\alpha \in \emptyset$), is usually not clearly articulated.

But actually Hoover was rediscovering the problem of passive observation as it was phrased in the early 1950s at the Cowles Commission, in a discussion about Herbert Simon's 1953 paper on causal ordering. To Simon, causal ordering had an "operational" meaning; it specified which variables would be affected by intervention at a particular point of the structure.

> We found that we could provide the ordering with an operational basis if we could associate with each equation of a structure a specific power of intervention, or "direct control." That is, any such intervention would alter the structure but leave the model (and hence the causal ordering) invariant. Hence, causal ordering is a property of models that is invariant with respect to interventions within the model, and structural equations are equations that correspond to specified possibilities of intervention. (Simon 1953, p. 66)

The operational meaning of causal ordering was pictured as follows:

> We suppose a group of persons whom we shall call "experimenters." If we like, we may consider "nature" to be a member of the group. The experimenters, severally or separately, are able to choose the nonzero elements of the coefficient matrix of the linear structure, but they may not replace zero elements by nonzero elements or vice versa (i.e., they are restricted to a specified linear model). We may say that they *control directly* the values of the nonzero coefficients. (Simon 1953, p. 65)

Turning any zero element into a nonzero one would change the causal structure. So causal ordering only had operational meaning within certain limits: "We must have a priori knowledge of the limits imposed on the 'experimenters'—in this case knowledge that certain coefficients of the matrix are zeros" (p. 65).

This emphasis on the importance of the distinction between zero and nonzero elements did not exist in an earlier discussion paper version. Compare the preceding quotation with the paragraph in the discussion paper version:

> We suppose a group of persons whom we shall call "experimenters." If we like, we may consider "nature" to be a member of the group. The experimenters, severally or separately, are able to choose without restriction the elements of a $n \times (n + 1)$ coefficient matrix. We may say that they *control directly* the values of these coefficients. (Simon 1950, p. 11)

It was probably Koopmans who pointed out to him the importance of a close connection between causal structure and the nonzero elements:

> I do not fully understand the role played by his "experimenters" who include nature, and who also set the coefficients which have prescribed values (zero). These experimenters seem to wield larger powers than human research workers do, and also larger powers than needed to give content to the notion of causal structure. A one-dimensional power of intervention in each equation is sufficient for the latter purpose. . . . The description of causal structures depends almost exclusively on information as to which coefficients are zero, which are not. (Koopmans 1951, pp. 2–3)

Simon did not give any indication how to attain this a priori knowledge. In fact, this is what here has been called the problem of passive observation—the problem of finding which coefficients are "ontologically" zero and which are not.

The forerunner of the concept of potential influence, the "strength of influence," that is, the standard deviation of a variable x times the regression coefficient, $a\sigma_x$, did not survive, either, even though one of the currently used test statistics, Student's t-statistic, to determine statistical significance seems to be a very similar concept. But it is not the same concept. When Frisch invented this concept the strength of influence the "probabilistic revolution" had not yet taken place in econometrics, The analysis, as we have seen, made no use of probability distributions. The variables were not yet considered to be stochastic, but determined instead by a causal mechanism.

In the early days of econometrics, when Frisch, Tinbergen, and also initially Haavelmo were aiming at an inductive science, they hoped that statistical significance would also mean economic significance. But Haavelmo came to the conclusion that statistical significance—dependent on actual variation—could not tell us about economic significance—telling us about potential variation, that is, what may happen.

A field science such as econometrics, where one cannot control and intervene, where Nature is the only experimenter but reluctant to inform us about the designs of her experiments, cannot be an inductive science based only on statistics and statistical methods and techniques to analyze them. We also need theory. But because theory is fundamentally incomplete—economics is an inexact science—we need additional sources of information. To know what these sources are, we have to leave econometrics and move to other disciplines, like medicine, where we will find more explicit accounts of these additional sources of knowledge. Koopmans and Haavelmo did not have them.

Notes

1. In her *Dappled World*, Cartwright also mentions a similar example of Otto Neurath's: "In some cases a physicist is a worse prophet than a [behaviorist psychologist], as when he is supposed to specify where in Saint Stephen's Square a thousand dollar bill swept away by the wind will land, whereas a [behaviorist] can specify the result of a conditioning experiment rather accurately" (Neurath quoted in Cartwright 1999, p. 27).
2. The concepts "economics" and "science" are of course terms that have to be considered in their historical context; they had meanings in the nineteenth century different from the ones they have now in the twenty-first. Nevertheless, I would like to show in this chapter that irrespective of the long period we are considering, the discussion of whether economics is an "exact science" has not changed considerably.
3. See Hausman (1992), in particular his chapter 8, for a detailed discussion of inexactness in economic theory. He also takes Mill's account of inexactness as a starting point but arrives at a different view on inexact laws than is proposed here.
4. With its 119 pages it can hardly be called an article. Actually it was Haavelmo's dissertation, which was not published as a separate book because of shortage of paper in wartime.

5. At that time, the board of directors consisted of P. Lieftinck, N. J. Polak, J. Tinbergen, and F. de Vries.
6. The Keynes-Tinbergen debate was about the epistemological reach of the at that time new method of econometric modeling, introduced to economics by Tinbergen. See next note.
7. Keynes's (1939) review article with Tinbergen's "reply" (1940) and Keynes's concluding "comment" (1940) constitute the Keynes-Tinbergen debate. Of interest here is that Keynes raised the same problem of completeness as Koopmans, so his point was addressed to someone who was very much aware of it: "Am I right in thinking that the method of multiple correlation analysis essentially depends on the economist having furnished, not merely a list of the significant causes, which is correct so far as it goes, but a *complete* list? For example, suppose three factors are taken into account, it is not enough that these should be in fact *veræ causæ*; there must be no other significant factor. If there is a further factor, not taken into account of, then the method is not able to discover the relative quantitative importance of the first three. If so, this means that the method is only applicable where the economist is able to provide beforehand a correct and indubitably complete analysis of the significant factors. The method is neither of discovery nor of criticism. It is a means of giving quantitative precision to what, in qualitative terms, we know already as the result of a complete theoretical analysis" (Keynes 1939, p. 560).
8. Bjerkholt (2005) suggests that Koopmans's visit to Oslo may have had a lasting influence ("a point of no return") on Haavelmo. Koopmans spent the autumn of 1935 at the University Institute of Economics and gave a series of lectures titled "Modern Sampling Theory," discussing Fisher's theory of estimation, and Neyman and Pearson's theory of hypothesis testing. Haavelmo (1938) reviewed Koopmans's (1937) dissertation.
9. Archives of the League of Nations, Palais de Nations, Geneva. I thank Pépin Cabo, Neil De Marchi, and Esther-Mirjam Sent for providing the copies of the files related to Tinbergen's project at the League of Nations.
10. Hendry and Morgan (1995, pp. 53–54) count 17 different kinds of tests. The one discussed here is number 7 on their list. The section of Tinbergen 1939a discussed here is also reprinted in Hendry and Morgan 1995, pp. 369–374.
11. Like Frisch, Haavelmo had the habit of coining new terms in every subsequent paper while actually discussing the same issues. Without being able to provide more textual evidence, I assume that Haavelmo used the term "average equation" (in the present paper) and the term "confluent equation" (in his other papers) as synonyms. In the same way, I assume that "momentary equation" is synonymous with "autonomous equation." Unlike the "confluent equations," the "autonomous equations" provide insight into the deeper structural equations. See chapter 3 of Boumans 2005 for a more detailed discussion of the concept of autonomy as used by Frisch and Haavelmo.
12. Section 10, pp. 33–39, in the 1941 version, and section 7, pp. 21–26, in the 1944 version. Both versions are exactly the same. Page numbers refer to the 1944 publication.
13. See Boumans 2005, p. 64, for the derivation of this expression.
14. ">>" means "sufficiently larger than," where "sufficient" is specified according to a specific metric, for example in terms of "statistical significant."
15. This is a Type A strategy, as it is called in chapter 2, where the partial derivatives are called sensitivity coefficients.
16. See my chapter "Autonomy" in Boumans 2005 for a detailed explanation of *ceteris neglectis* environments.
17. A similar point is made in Chang's *Inventing Temperature* (2004), where he calls the process of "science founded on measurement founded on the definition of measurands founded on science" an "epistemic iteration," see chapter 3.
18. By distinguishing between the problem of passive observation and the problem of identification, this chapter deviates from Duo Qin's (1993, pp. 104–105) historical account, which discusses the problem of passive observation only in the context of identification.

5

Clinical Judgment

Where is the wisdom we have lost in knowledge?
Where is the knowledge we have lost in information?

—T.S. Eliot, *The Rock*

5.1. Introduction

In the preceding chapters it was argued that for reliable measurement, mechanical procedures are not sufficient, and thus an additional source of knowledge is needed. This source of knowledge has been given various names: "expert judgment," "scientific judgment," "skilled judgment," "trained judgment," and "Type B judgment," without having said much about its meaning. This chapter aims to clarify what kind of judgment is referred to and what it is not.

In terms of selling ideas, this is not the best time to write a book on measurement in which it is claimed that expert judgment is indispensable for acquiring reliable measurement. Even worse is that I would have liked to use the term "clinical judgment" in its title. I did not do so, because it would confuse potential readers by making them think that this book is only about measurement in medicine. But the reason I was considering the term "clinical judgment" as a key concept for this book is that for explaining what it entails to arrive at reliable measurements outside the laboratory, the term "clinical judgment" points at one of its main problems: finding an optimal balance between objectivity and subjectivity.

The reason the adjective "clinical" is particularly helpful in explaining the purpose of this book is because of the double meaning of the term "clinical." On the one hand it means "coldly detached and dispassionate" (*Oxford English Dictionary*), so to pertains to a judgment that is unemotional, scientific, objective, analytic, impersonal, clean, disinterested, and emotionless. On the other hand, it means "of or pertaining to the sick-bed" (*Oxford English Dictionary*), and in a dictionary specialized in medical terms it pertains to or is founded on "actual observation and treatment of patients, as distinguished from theoretical or experimental" (*Online Medical Dictionary*). This second meaning becomes even more significant when taken as an adjective to modify "judgment," that is, in "clinical judgment": "the critical decisions made on the basis of scientific observations but with the added skill provided by long experience of similar cases" (*Online Medical Dictionary*). In current debates in relation to evidence-based medicine, "clinical" (as an adjective modifying judgment) even has a meaning opposite to the first meaning I have given: biased, preconceived, prejudiced, and subjective.

The evidence-based-medicine movement emerged in the early 1990s, and its approach was propagated by the Evidence-Based Working Group (1992).[1] Evidence-based medicine was presented as a "new paradigm for medical practice":

> Evidence-based medicine de-emphasizes intuition, unsystematic clinical experience, and pathophysiologic rationale as sufficient grounds for clinical decision making and stresses the examination of evidence from clinical research. Evidence-based medicine requires new skills of the physician, including efficient literature searching and the application of formal rules of evidence evaluating the clinical literature. (Evidence-Based Working Group 1992, p. 2420)

One of the "foundations of the paradigm shift" is the increasing dominance of the randomized clinical trial. Current evidence-based-medicine assessments make use of an "evidence hierarchy," often called "levels of evidence," in which higher levels of evidence are regarded as of higher quality than lower levels. A typical evidence hierarchy puts double-blinded randomized clinical trials at the top and expert opinion at the bottom (see, for example, figure 5.1: "Expert reports of their clinical experience should be explicitly labeled as very low quality evidence, along with case reports and other uncontrolled clinical observations" [Guyatt et al. 2008, p. 925]). The rationale for the evidence hierarchy is that higher levels of evidence are thought to

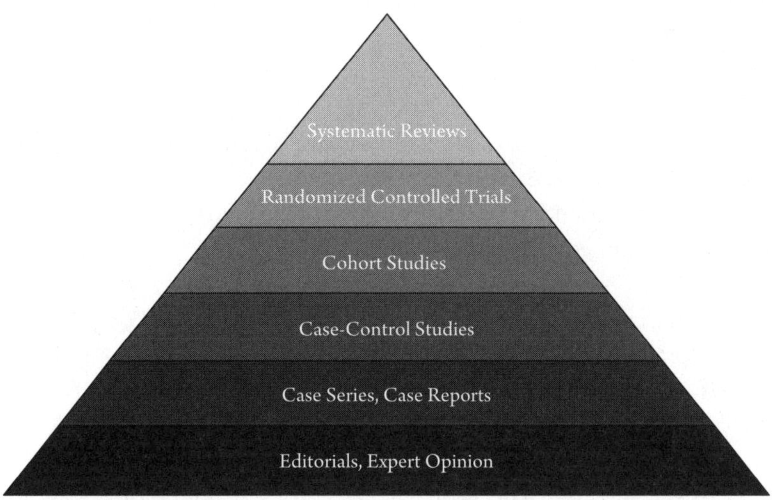

Figure 5.1 Levels of Evidence. Source: Information Services Department of the Library of the Health Sciences-Chicago, University of Illinois at Chicago, 2006.

avoid biases that are present in the lower levels of evidence. Note that the pyramid in figure 5.1 uses the term "expert opinion," and Guyatt and coauthors (2008) use the term "expert report" instead of "clinical judgment."

The probable reason for not putting "clinical judgment" at the bottom of the evidence hierarchy but "expert opinion" instead is that on the one hand evidence-based medicine is explicitly opposed to experts and expertness, most explicitly and provocatively voiced by the "father of evidence based medicine," David L. Sackett:

> It is then dawned on me that experts like me commit two sins that retard the advance of science and harm the young. Firstly, adding our prestige to our opinions gives the latter far greater persuasive power than they deserve on scientific ground alone. Whether through deference, fear, or respect, others tend not to challenge them, and progress towards the truth is impaired in the presence of an expert. . . . new ideas and new investigators are thwarted by experts, and progress toward the truth is slowed. . . . Is redemption possible for the sins of expertness? The only one I know that works requires the systematic retirement of experts. . . . But there are still more experts around than is healthy for the advancement of science. (Sackett 2000, p. 1)

On the other hand, it is still not settled yet within evidence-based medicine whether one can also abandon clinical judgment. Take, for example, a recent debate on evidence-based medicine and clinical judgment that appeared in two issues of the *Journal of the American College of Cardiology*. This debate started with Alexandre C. Pereira and coauthors (2006), who describe clinical judgment (in chronic illness) as involving "knowledge of the natural history of the disease, the ability to assess the validity of therapeutic claims, and a means of applying what is known about the individual patient" (p. 948), and conclude their article by stating,

> Our data, together with that of others, are a reminder that physician judgment remains an important predictor of outcomes, even in a time when evidence-based medicine is considered the gold standard of medical practice. (Pereira et al. 2006, p. 953)

In an editorial comment on this article, Ori Ben-Yehuda (2006, p. 954) observes that the randomized clinical trial has become "the apotheosis of our endeavor to determine the efficacy of different therapies. . . . The overall success of the modern clinical trial has led to a great emphasis on what has been termed evidence-based medicine." Acknowledging this success of randomized clinical trial design, in particular with respect to complex decisions, he also notes its limitations: "Randomized clinical trials, whether of a pharmacologic or a mechanical intervention, are based on the assumption that patients in the different arms being compared have similar baseline characteristics" (p. 954). But "the physician as well as patient preference may include a host of factors beyond those included in clinical trial criteria and that may have biologic significance" (p. 954). These observations led Ben-Yehuda to raise the rhetorical question:

> Is there room, therefore, in this evidence-based era for individual physician and patient judgment? Are outcomes incorporating such judgment better than the reliance on evidence-based clinical trial data? (Ben-Yehuda 2006, p. 954)

In a letter to the editor, Ganesan Karthikeyan (2007) responded to Pereira and coauthors' article and Ben-Yehuda's commentary by interpreting their contributions as arguing against evidence-based medicine:

> Detractors of evidence-based medicine tend to imbue "clinical judgment" with an aura, which barely falls short of the divine, by attributing intangible powers to clinicians. This view of clinical judgment is more about the clinician than about judgment. In reality, individuals, clinicians, or otherwise, are swayed more by anecdotal experience; as a result, they are more prone to systematic errors while making judgments under situations of uncertainty. Evidence from clinical trials, if anything, adds objectivity, reduces bias, and refines a clinician's ability to make decisions. (Karthikeyan 2007, p. 1012)

To support his claim about clinicians being "more prone to systematic errors," he refers to Amos Tversky and Daniel Kahneman's (1974) *Science* article "Judgment under Uncertainty: Heuristics and Biases" (on which more subsequently).[2]

In their "Reply," Pereira and Whady Hueb (2007, p. 1012) emphasize that "far from a mystical definition," clinical judgment is a combination of both objective and subjective information: "the result of a complex equation that takes into account objective data from biochemical tests, imaging studies, and a patient's history. It also uses subjective information acquired by the physician over the course of the patient-physician relationship." Because of the broad spectrum of variables, clinical, demographic, angiographic, and biochemical, that are being used, no simple judgments can easily be made, so medical guidelines, randomized controlled trials, and cost-effective algorithms are "invaluable for practicing medicine and in helping the decision-making process. Nevertheless, we should not forget that a physician's judgment is what processes and consolidates all this information" (p. 1013).

Ben-Yehuda's (2007) "Reply" is very much in the same vein: his main point was that "the complexity of clinical decision process as well as the uniqueness of each individual patient may not always be adequately captured in our evidenced-based criteria" (p. 1013):

> No amount of clinical trial data can ever capture the almost infinite variables involved in the complex biology of health and disease. In addition, the somewhat arbitrary cutoffs employed in data analysis add additional limitations. (Ben-Yehuda 2007, p. 1013)

This debate in medicine illustrates nicely one of the central problems with respect to "measurement outside the laboratory." When the measurand is simple, usually there are no additional problems compared to measurement in a laboratory. But most phenomena are not simple; an "almost infinite" number of factors are involved of which not all can be measured or are even known. Moreover, whatever mechanical measurement procedure is developed, it always has to assume "similar baseline characteristics," that is, it always has to abstract from idiosyncratic circumstances, which have repercussions for the reliability of the measurement of a field phenomenon. When pure objectivity is impossible, we need to accept subjectivity to complement the incomplete objective knowledge. The question is not how to exclude subjective judgment, but rather where do we allow it, how much, and in what sense?

It is generally acknowledged that "data underdetermine theory," but this applies equally to method and procedure. Data will not tell you how they should be treated. The choice of the "calculus" (discussed in chapter 3) is based on the assumed level of reliability each treatment can provide, but reliability has different facets—precision, accuracy, standardization, stability, certainty, unbiasedness—that usually conflict with each other. These different facets express different epistemic values. Someone has to make a choice; there exists no "golden rule" that without intervening expert judgment can be applied mechanically. Measurement is the combination of mechanical procedures and trained judgment. This is nicely illustrated by Peter Sydenham in his (1979) book on measuring instruments by a figure picturing a measuring system; see figure 5.2. In this almost completely mechanized measuring system, human judgments—represented by the two human figures here—are considered to be necessary links of the complete measurement chain.

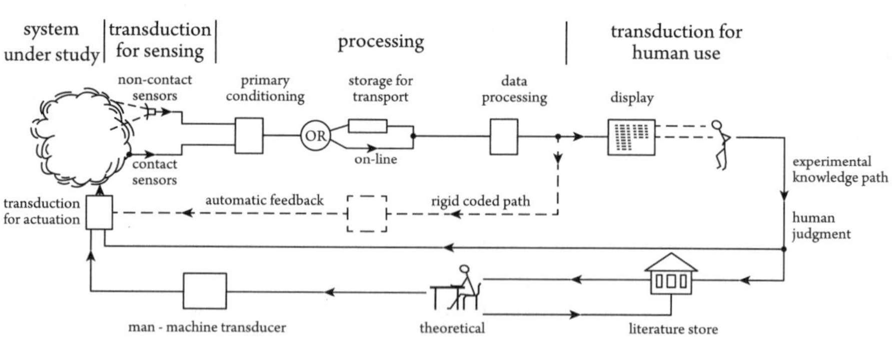

Figure 5.2 Human Judgment as Part of a Measuring System. Source: Sydenham 1979, p. 19.

This picture shows a system in operation, so seemingly only a few human beings are needed. But a picture of this system in its construction phase would have been crowded by human figures. Every part of as well as the system as a whole is designed, planned, made, and installed by human figures. Subjective judgment is everywhere.

Chapter 6 will discuss how subjective judgment can be combined with objective procedures. Here, in this chapter, we will focus on the question of what kind of judgment is needed as a complement to the mechanical procedures.

5.2. Judgment

> You see, ladies and gentlemen, and above all Your Imperial Majesty, with a real nightingale one never knows what to expect, but with this artificial bird everything goes according to plan. Nothing is left to chance. I can explain it and take it to pieces, and show how the mechanical wheels are arranged, how they go around, and how one follows after another.
>
> —"The Nightingale," Hans Christian Andersen, 1843

A measurement is the outcome of a combination of mechanical objectivity and "considered judgment." According to Julian Reiss (2014), who borrows this term from Catherine Elgin (1996), "considered judgment" about a hypothesis involves

> taking into account all the evidence relevant to the assessment of the hypothesis, which requires judgments about relevance, about the quality of the evidence, about weighing difference pieces of evidence, about the amount of evidence needed to accept or act on the hypothesis, about whether or not additional evidence should be sought in the light of the cost of doing so, and so on. Many of these judgments do not follow strict rules and are therefore "subjective" in the eyes of some. (Reiss 2014, p. 138)

Judgment is defined here in the Kantian sense, that is, as the application of a general rule to a particular case, a universal to an instance of that universal. But it is not a strict deduction of conclusions from premises, it has cognitive content, and there is expertise in it, that is, real experience with the particulars to which it is applied. Judgment plays a role "anywhere that conclusions depend in significant part on grasping the features of complex and unreproducible particular cases" (Fleischacker 1999, p. 9).[3]

Immanuel Kant ([1892] 1914) distinguishes between two kinds of judgment: reflective and determining judgment. If the general (the rule, the principle, the law) is given, then the judgment that subsumes the particular under it is "determining." If, however, only the particular is given for which the general must be found, then the judgment is merely "reflective."

To develop a "third concept of liberty" that is based on Kant's aesthetic judgment, Samuel Fleischacker (1999) argues that it is important that judgment leave enough room for "reasonable disagreement" (p. 9). Judgment is defined as the conclusion of a train of thought "where the interpretation of particular cases is essential to that train of thought," and "the conclusions to which one comes will always be open to further debate" (p. 9). As a result, the mathematical or logical derivation from given premises or scientific laws is not a judgment.

Fleischacker argues that the way we come up with rules (reflective judgment) and the way we apply them (determining judgment) are different sides of the same operation:

> What we do in reflective judgment is reinterpret an object that we feel we have hitherto insufficiently or inaccurately conceptualized. We open up conceptual applications that we previously took to determine the object; we shift the object into a different set of intellectual boundaries. We may also, thereby, shift the boundaries of our intellectual sets, our concepts, themselves. (Fleischacker 1999, p. 27)

So, according to Fleischacker, reflective judgment not only consists in a play between concepts and intuitions, but also participates in an interplay with determining judgment as well: Concepts have a definite meaning insofar as we have a definite system of scientific determining judgments, but such system must be constantly scrutinized for evidence as to the facts, and "that means that our determining judgments, and the concepts they define, must always stand open to being reinterpreted, reshaped into a new system, by reflective judgment" (p. 28).

> As Kant characterizes them, reason, understanding, and imagination are lonely and silent processes, shared with others and informed by others only via their coming together in judgment. We might say that reason and understanding are too sure of themselves, too definitive, to provide anything worth discussing, while imagination is too inchoate, too *in*definite to make conversation possible. What we have already placed into a rational theory or category of a theory [read: model] is beyond interesting discussion; what we have merely sensed, with all the peculiarities of our individual capacities for sensation, is not yet expressible in linguistic terms. Only judgment is simultaneously formed enough to be discussible and indeterminate enough to be worth discussing. (Fleischacker 1999, p. 31)

Judgment presupposes imagination, while inferences from a model can be done mechanically, aptly expressed by Henk Don: "An economic model has a good memory but little imagination."[4]

But if the imagination does not come from the model, where does it come from? From humans obviously. But not just anyone is supposed to make an appropriate judgment with respect to any domain. Someone who is invited to make such judgments is supposed to be an "expert" on the relevant domain, that is, someone who is able to make a "scientific judgment," that is, who has the scientific training and reputation, and the experience with the subject and with the behavior and properties of the relevant materials and instruments. The next chapter will discuss the selection of appropriate experts in more detail.

5.3. Rationality

> He's right and he's right?
> They can't both be right.
> You know, you are also right.
>
> —*Fiddler on the Roof*[5]

Once one has accepted that judgments have an indispensable role in science, accounts can be developed to ensure that judgments are "rational" to eliminate their subjectivity. But as will be shown here, in these accounts a rational judgment is usually considered to be the optimal solution of a real-life judgment problem that is modeled in a specific way. Each model imposes a specific rational judgment as its optimal solution. The term "solution" is chosen purposefully to indicate that once a model and data input have been chosen, a "rational" judgment is the result of an (often probabilistic) calculation. So the "rationality" of a judgment depends on the particular specification of the model for a specific problem, and is therefore model-based. Actually, a rational judgment in this sense is not a judgment in the Kantian sense. The essential difference between a rational judgment and a clinical judgment (see above) is that the first is completely defined by the relevant model and the latter refers to knowledge that goes beyond that model (e.g., imagination) and is needed to build the model.

Formulated in this way, and this is how "rational judgment" is often understood, rational judgment is equivalent to measurement as defined by the axiomatic theory of measurement (see chapter 2, section 2.2).[6] Although one could say that the difference between rational judgment and measurement is that judgment refers to an assessment of a human agent while measurement refers to an assessment of a measuring system, the latter nevertheless includes trained judgment beside mechanical objectivity (see, for example, figure 5.2). The difference between rational judgment and measurement defined by the axiomatic theory of measurement, however, is only linguistic. For example, like a measurement, a judgment can be "biased." Although a judgment that is "biased" is also called "irrational," the two terms have the same meaning.[7]

Although rational judgment refers to human judgment, there is an accumulation of literature showing that human judgments are not rational (see, for example, in the evidence-based medicine literature previously discussed). The most influential contributions to this literature over the past four decades are the publications of Amos Tversky and Daniel Kahneman, particularly their jointly written *Science* article (1974) "Judgment under Uncertainty: Heuristic and Biases," and the (1982) volume jointly edited with Paul Slovic with the same title, in which the *Science* article appeared as the first chapter.

By judgment Tversky and Kahneman (1974, p. 1124) meant the assessment of probability.[8] They argued that people rely on a limited number of heuristic principles that reduce the complex tasks of assessing probabilities and predicting values to simpler judgmental operations. In general, these heuristics are quite useful, but sometimes they lead to severe and systematic "errors," which Tversky and Kahneman called "biases." They describe three heuristics with accompanying biases: "representativeness," "availability," and "adjustment and anchoring."

The representativeness heuristic is applied for probabilistic questions, such as What is the probability that process B will generate event A? In answering this question, Tversky and Kahneman (1974) claim that the probabilities are evaluated by the degree to which A is "representative" of B, that is, by the degree to which A "resembles" B. One of the reasons that this heuristic leads to biases is its "insensitivity to prior probabilities of outcomes." This bias is better known as the "base-rate fallacy." This fallacy will be discussed in section 5.4.

Another bias, the "gambler's fallacy," of the representativeness heuristic is the "misconception of chance": people expect that a short sequence of events generated by a random process will represent similar characteristics as a long sequence generated by the same random process. This fallacy will be briefly discussed in section 5.5.

The availability heuristic is the assessment of the probability of an event by the ease with which instances or occurrences can be brought to mind. Anchoring is the judgmental heuristic when estimates are made by starting from an initial value that is adjusted to yield the final answer. The bias results from "typically insufficient" adjustments. This fallacy will be discussed in section 5.6.

When discussing the different heuristics and biases, Tversky and Kahneman (1974) note that reliance on heuristics and prevalence of biases are not restricted to "naive subjects" and "laymen"; they also include, for example, experienced research psychologists: "Experienced researchers are also prone to the same biases—when they think intuitively" (p. 1130). This sentence became one of the most quoted in the evidence-based literature, but by leaving out the last part: Experts are prone to biases, full stop.

The biases Tversky and Kahneman discussed in their 1974 article were fallacies they already had detected in several experiments run by themselves and others, but the article initiated a huge industry of experiments where in several different ways it was shown that people—laymen and experts—again and again make biased judgments. The edited 1982 volume *Judgment under Uncertainty* resulted from these experiments.

The problem, however, is that whether a judgment (or measurement) is biased or unbiased depends on the model or (formal) representation of the judgment problem setting and the judgment problem itself. For probabilistic assessment models, this presupposes drawing on assumptions that frame the problem as accurately as possible in three dimensions: demarcation of the sample space, definition of the target probability, and construction of the information structure. In probability theory, a sample space is defined as the set of all possible outcomes. The target probability is the probability of the targeted outcome, for example, maximum utility. And the information structure is the set of conditional probabilities telling what the probability of some signal is given the occurrence of some event. Which outcomes and signals are possible and what precise probability is targeted are all specific interpretations of the problem that determine the model assumptions.

But modeling is not an obvious task, and one can have different interpretations of what the problem entails, particularly in the case of a real-life problem. Moreover, there is no higher authority that decides which model is the most adequate representation of a real-life problem. Experts may differ on interpretations, and so can come up with different solutions of how to assess rationality. As a result, different models may impose different, even contradictory, rational judgments, creating "anomalies" or "paradoxes."[9]

This assertion that different models of a problem lead to different rational solutions will be demonstrated by three cases. These three cases show that "rational judgment" is constrained to a prior selected model, whereas "considered judgment" considers also the choice of a model and the selection of evidence.

Moreover, this assertion about the close relation between "model" and "rational solution" also has an important consequence for the design of experiments. An experiment on a judgment problem is a designed setting according to an experimenter's framing of that judgment problem. For an experiment, the experimenter is the expert who knows the accurate representation of the problem (and when designing the experiment has verified whether this representation is indeed accurate). In other words, the difference between a real-life situation and an experiment is that in an experiment the three dimensions I have mentioned, sample space, target probability, and information structure, are known to the experimenter. But although we can assume that the experimenter has an accurate model of the experiment, one cannot simply assume this is also the case for the subjects who participate in this experiment. Subjects may have a different model in mind when making their decisions and so make irrational decisions according to the experimenter's model but not from their own perspectives, which leads to the experimenter's astonishment: subjects persist in being irrational even after the subject has been enlightened by the experimenter about his or her fallacy.

This problem of the possibility of having two different models (the experimenter's and the subject's model) in an experiment is closely related to a problem already noted in psychology in the early 1970s by Martin T. Orne (1973) and labeled by John G. Adair (1984) as the "two-experiment problem." According to Orne, in any study there are potentially two experiments: the experiment designed by the experimenter and the experiment perceived and participated in by the subjects. Because it is the subject's perception rather than the experimenter's that will determine how the subject responds, "it is essential for the investigator to understand how a particular experimental situation is perceived by the subject in order to draw sensible inference from the subject's responses" (Orne 1973, p. 158).

Discussing the "artifact crisis" in social psychology, Adair (1991) called the two-experiment problem "our greatest problem":

> When subjects react to an experiment that is different from the one the experimenter intended, artifact results and erroneous conclusions are drawn. . . . Failure to understand how the task is revealed to the subject may have resulted in a tendency to over-infer irrationality and bias in human judgment. (Adair 1991, pp. 447–448)

Although human subjects manifest some flawed reasoning and misjudgments, the consistent attribution of systematic and persistent biases to subjects, called "intuitive statisticians," may also suggest a dispositional bias on the part of the investigators, that is, the unawareness that the subject and experimenter may have two different perceptions of the experiment.

The acknowledgment of the two-experiment problem in psychology and cognitive science led to various kinds of innovations, like greater attention to instructions to subjects, continued improvement in techniques for assessing subjects' cognitions, and increased sensitivity to subjects' phenomenological experiences as research participants. What, however, is not considered explicitly is the possibility that a problem can be framed in two or more different models and what the consequences of this are for studying rationality. I would like to call this the two-model problem, in analogy with the two-experiment problem, but I also need to indicate that this is a different kind of problem. For example, the two-experiment problem entails the framing effect. Framing effects refer to the possibility that alternative ways of posing an identical problem may affect an agent's choices (see, e.g., Camerer 1995). So, in the framing case, the problem is defined equally for different framings, but in the two-model case the definition of the problem and so its solution may differ for each model.

This remaining part of this chapter will discuss the two-model problem for three cases. The first case is a comparison of a medical judgment model with a statistical judgment model leading subsequently to two different assessments of rationality in medical practice. The second case is a comparison of two models on the existence of the "hot hand" in basketball and the subsequent rational assessment of a coach to keep a player in the game or not. The third case is the notorious Monty Hall (three doors) paradox, which led to an ongoing debate in the mathematical statistical literature on how to model this paradox.

5.4. Judgment in medicine

> "We are coming now rather into the region of guesswork," said Dr Mortimer.
> "Say, rather, into the region where we balance probabilities and choose the most likely. It is the scientific use of the imagination, but we have always some material basis on which to start our speculations."
> —Arthur Conan Doyle, *The Hound of the Baskervilles*

A classic example of a "base rate fallacy" is the "Harvard Medical School Test." It appeared that, when a laboratory test result is given, physicians do not take account of the base rate, or pretest probability, to reach a clinical decision. This Harvard Medical School Test, carried out by Ward Casscells, Arno Schoenberger, and Thomas Graboys (1978), was a small survey to obtain some idea of how physicians interpret a laboratory result.

> We asked 20 house officers, 20 fourth-year medical students and 20 attending physicians, selected in 67 consecutive hallway encounters at four Harvard Medical School teaching hospitals, the following question: "If a test to detect a disease whose prevalence is 1/1000 has a false positive rate of 5 per cent, what is the chance that a person found to have a positive result actually has the disease, assuming that you know nothing about the person's symptoms or signs?" (Casscells, Schoenberger, and Graboys 1978, p. 999)

Using Bayes' theorem, the "correct" answer should be 2%.[10] The result of this test was that only 11 out of 60 participants gave this answer. The most common answer, given by 27, was 95%. The average of all answers was 55.9%, "a 30-fold overestimation of disease likelihood" (p. 1000).

Discussing these results, Casscells, Schoenberger, and Graboys observe that, despite the fact that probabilistic reasoning has been presented in prominent clinical journals for a decade, "in this group of students and physicians, formal decision analysis was almost entirely unknown and even commonsense reasoning about the interpretation of laboratory data was uncommon" (p. 1000). This problem, however, was considered to be remediable by practical instruction in the theory of test interpretation.

Four years later a similar result was obtained by David M. Eddy (1982), which was published in *Judgment under Uncertainty*, edited by Kahneman, Slovic, and Tversky. Eddy discusses a more specific case of deciding whether to perform a biopsy on a woman who has a breast mass that might be malignant. Specifically, he studied how physicians process information about the results of a mammogram, an X-ray test used to diagnose breast cancer.

The prior probability, $Pr(ca)$, "the physician's subjective probability," that the breast mass is malignant is assumed to be 1%. To decide whether to perform a biopsy or not, the physician orders a mammogram and receives a report that in the radiologist's opinion the lesion is malignant. This is new information and the actions taken will depend on the physician's new estimate of the probability that the patient has cancer. This estimate also depends on what the physician will find about the accuracy of mammography. This accuracy is expressed by two figures: "sensitivity," or true-positive rate $Pr(+|ca)$, and "specificity," or true-negative rate $Pr(-|benign)$. They are respectively 79.2% and 90.4%. Applying Bayes' theorem leads to the following estimate of the posterior probability: 7.7%.[11] In an "informal sample" taken by Eddy, most physicians (approximately 95 out of 100) estimated the posterior probability to be about 75%.

When Eddy asked the "erring" physicians about this, they answered that they assumed that the probability of cancer given that the patient has a positive X-ray, $Pr(ca|+)$, was approximately equal to the probability of a positive X-ray in a patient with cancer, $Pr(+|ca)$.

> The latter probability is the one measured in clinical research programs and is very familiar, but it is the former probability that is needed for clinical decision making. It seems that many if not most physicians confuse the two. (Eddy 1982, p. 254)

According to Eddy, it is not only the participating physicians who were erring, but a review of the medical literature on mammography reveals a "strong tendency" to equate both probabilities, that is, to equate $Pr(ca|+) = Pr(+|ca)$. He saw that erroneous probabilistic reasoning was widespread among practitioners, and therefore focusing on improving this kind of reasoning would have an important impact on the quality of medical care:

> The probabilistic tools discussed in this chapter have been available for centuries. In the last two decades they have been applied increasingly to medical problems . . . , and the use of systematic methods for managing uncertainty has been growing in medical school curricula, journal articles, and postgraduate education programs. At present, however, the application of these techniques has been sporadic and has not yet filtered down to affect the thinking of most practitioners. As illustrated in

this case study, medical problems are complex, and the power of formal probabilistic reasoning provides great opportunities for improving the quality and effectiveness of medical care. (Eddy 1982, p. 267)

Eddy's explanation for this specific judgmental bias is that most practitioners fail to reason probabilistically. Today, probabilistic reasoning means in most cases Bayesian reasoning. And so correct Bayesian reasoning is supposed to lead to an "unbiased judgment." In the literature whenever "biases" are discussed, it is not made clear what in each case is meant by "bias." The term often alludes to probabilistic bias. In the appendix to this chapter, however, it is shown that correct Bayesian reasoning is not necessarily equal to probabilistic unbiasedness.

Thus, from the outside, it looks very much that "many physicians make major errors in probabilistic reasoning, and that these errors threaten the quality of medical care" (Eddy 1982, p. 249). Considering the issue from the perspective of medical practice, however, one arrives at a different view of physicians' decision-making. What Casscells, Schoenberger, Graboys, and Eddy overlooked was that a test is not like an innocent drawing from an urn with colored balls. Tests can be painful or risky, so a clinician only asks for a test after a well-considered evaluation of reliability, value, and risk. The decision to ask for a test is usually based on an evaluation using Stephen G. Pauker and Jerome P. Kassirer's (1980) threshold model. Pauker and Kassirer developed this threshold model as a tool to make this decision to test or not objective, and not to be a "toss-up." In their article they describe a model that uses two thresholds to aid physicians in making clinical decisions:

- A "no treatment/test" threshold, T_t, which is the disease probability at which the expected utility of withholding treatment is the same as that of performing a test
- A "test/treatment" threshold, T_{trx}, which is the disease probability at which the expected utility of performing a test is the same as that of administering treatment

The decision "not to treat," "to test," or "to treat" is determined by the pretest disease probability and both thresholds. The best clinical decision for probabilities below the "no treatment/test" threshold T_t is to refrain from treatment; for probabilities above the "test/treatment" threshold T_{trx}, the best decision is to administer treatment. When the pretest disease probability lies between the thresholds, the test result could change the probability of the disease enough to alter the decision, so the best decision would be to administer a test. As such, for clinical decision-making, estimates of base rates are crucial.

It is noteworthy to see that Pauker and Kassirer (1980), though referring to Tversky and Kahneman (1974), make, unlike Tversky and Kahneman, a distinction between expert opinions and opinions of those outside the medical domain: "Studies in nonmedical domains show that people have biases and often make inaccurate estimates and that training improves the reliability of such estimates" (p. 1112). According to Pauker and Kassirer, the sort of studies Tversky and Kahneman are referring to rely on "simple tests in which an actual probability is known (for example, the number of various coloured balls in an urn)," whereas in medicine a prevalence "represents a belief or opinion for which no actual or true value exists" (p. 1112). Moreover, it appears that physicians make probability estimates with "reasonable reliability" (p. 1112). When published data on probabilities are not specific enough, the "opinions of experts" are needed and used.

From the threshold model already described, test criteria can be inferred. In the first instance, according to Barbara Scherokman (1997), tests that do not change the probability of disease enough to cross the threshold probability T_{trx} are not useful and should not be ordered. This means that when the pretest disease probability lies between the thresholds and we have a positive test result, the post-test disease probability should lie above the test/treatment probability: $Pr(P|+) > T_{trx}$.

This is in fact a weak criterion, because it implies that the disease at least should (causally) influence the test result: $Pr(+|P) > Pr(+)$.[12] In statistics, an event A is called "independent" of event B if $Pr(A|B) = Pr(A)$. In probabilistic accounts of causality, it is crudely stated that B causes A if $Pr(A|B) > Pr(A)$. So the preceding requirement only excludes tests like flipping a coin, a "toss-up."

A stronger test requirement is that it should be "most informative." A test is most informative when its predictive values, $Pr(P|+)$ and $Pr(A|-)$, are optimal. As I have already mentioned, these values depend on the pretest probabilities. It can be shown that both predictive values are optimal when $Pr(P) = 1/\left(\sqrt{LR(+)LR(-)} + 1\right)$, where $LR(+) = Pr(+|P)/Pr(+|A)$ is the likelihood ratio for a positive test result, and likewise $LR(-)$ is the likelihood ratio for a negative test result.[13]

Usually the test characteristics' "sensitivity" and "specificity" are about equal, which means that the optimal pretest probability is about 50%.[14] Generally, it is expected that a test is

> most informative when the pre-test probability of disease is between 40% and 60%. In other words, a diagnostic is most useful and changes the pretest probability of disease if the patient is believed to have a 50:50 chance of having the disease. (Scherokman 1997, p. 6)

These demands on tests with respect to accuracy and applicability shed new light on the interpretation by physicians of clinical laboratory results. First, assume that the condition for using the test is optimal: $Pr(P) \approx 0.5$, so $Pr(A) = 1 - Pr(P) \approx 0.5$. When sensitivity and specificity are about equal, then $Pr(+) \approx 0.5$.[15] So if physicians assume that a test is used for optimal conditions, there is no question of base rate fallacy, because

$$Pr(P|+) = \frac{Pr(+|P)}{Pr(+)} Pr(P) \approx Pr(+|P).$$

Second, let us take Eddy's figures: $Pr(+ \mid P) = 79.2\%$ and $Pr(- \mid A) = 90.4\%$, and assume that $40\% < \Pr(\mathrm{P}) < 60\%$, then $37.44\% < Pr(+) < 51.36\%$, and so

$$84.6\% < Pr(P \mid +) < 92.5\%.$$

Most physicians estimated the post-test probability to be about 75%.

And finally, the Harvard Medical School Test figure, $Pr(+ \mid A) = 5\%$, leads to even higher post-test probabilities, when the prevalence is between 40% and 60%:

$$93\% < Pr(P \mid +) < 95\%.$$

Recall that most common answer, given by 27 out of 60, was 95%.

Physicians are trained not to ask for diagnostic tests when prevalences are too small (or too large).[16] Faced with test results they might have assumed—due to their training—that the test was performed for the right conditions. So they might have developed a heuristic to read the sensitivity and specificity as predictive values. Seen from this perspective, the physician's high estimates of the post-test probabilities in the case of the Harvard Medical School Test and in Eddy's test are not biased, but "rational."

In a discussion of the Monty Hall problem (see section 5.6) in the *American Statistician*, one of the debaters, John P. Wendell, referred to the Harvard Medical School Test in his argumentation and arrived at a conclusion similar to the one here. It is thus worth quoting at length:

> This answer of 2% apparently assumes that everyone in the population, whether they have the disease or not, has an equally likely chance of receiving the test and that the false negative rate is zero. . . . Neither of these assumptions is stated or clearly implied in the problem. Stating "you know nothing about the person's symptoms or signs" is not the same as stating that the test has an equal chance of being administered to people in the population, even if that was the intent of the phrase. The medical students and staff that were given this question would know full well that patients having a disease are almost always more likely to have a test for their disease administered to them than the general public. . . . The majority response of 95% is consistent with the assumption that persons having the test applied to them have a 50% chance of actually having the disease. . . . Certainly these assumptions are more reasonable than those needed to support the 2% answer. Perhaps this illustration shows not that medically trained people don't understand probability but that some statisticians don't understand medicine. (Wendell 1992, p. 242)

To this comment, Richard G. Seymann (1992, p. 242) responded that to "know nothing about the person's symptoms or signs" is not an instruction to assume random testing, but "a clear instruction to disregard all other information, biases, or prejudices we might have":

> One must ask where a 50% prior comes from. . . . The argument for a 50% prior, though perhaps understandable in other circumstances, here results in the fabrication of a new prior and the dismissal of a vital piece of explicit information. (Seymann 1992, p. 242)

The issue here of course is that what is a "vital piece of explicit information" for a statistician is not necessarily the same as for a physician.

To illustrate this difference between what kind of information is considered to be "vital" for a practitioner and a statistician, take one of the "fables" of Russell L. Ackoff:

> In a conversation with one of my colleagues I was asked how I would go about determining the probability that the next ball drawn from an urn would be black if I knew the proportion and number of black balls that had previously been drawn. He told me that the urn contained only black and white balls. I replied that I would first find out how the urn had been filled. "No," he said, "that is not permissible." "Why?" I asked, "Certainly you have such information." "No, I don't," he replied. "Then how do you know the urn contains only black and white balls?" I asked. "I have it on good authority," he answered. "Then let me talk to that authority," I countered. In disgust he told me to forget about the whole thing because I clearly missed the point. I certainly did. (Ackoff 1974, p. 89)

The moral of this fable is that the ability to solve a textbook exercise is not equivalent to the ability to solve a real-world problem. Textbook exercises are usually formulated so as to have only one correct answer and one way of reaching it. Real-world problems have neither of these properties. An essential part of problem solving, according to Ackoff, lies in determining what information is relevant and in collecting it.

By discussing six problems in reasoning with probabilities, so-called "teasers," Maya Bar-Hillel and Ruma Falk (1982) show that the way we model a problem is strongly dependent on the way the information was obtained.

> The kind of problem in which the conditioning event does turn out to be identical to what is perceived as "the information obtained" can only be found in textbooks. Consider a problem which asks for "the probability of A *given* B." This nonepistemic phrasing sidesteps the question of how

> the event B came to be known, since the term "given" supplies the conditional event, by definition. . . . Outside the never-never land of textbooks, however, conditioning events are not handed out on silver platters. They have to be inferred, determined, extracted. In other words, real-life problems (or textbook problems purporting to describe real life) need to be *modeled* before they can be solved formally. And for the selection of an appropriate model (i.e., probability space), the way in which information is obtained (i.e., the statistical experiment) is crucial. (Bar-Hillel and Falk 1982, pp. 120–121)

Bar-Hillel and Falk (1982) emphasize that a probability space for modeling verbal problems should allow for the representation of the given outcome and the statistical experiment that yields it. They illustrate how different scenarios for obtaining some information yield different solutions. In other words, the way one models a problem is strongly dependent on how the information is obtained. Different ways of obtaining the same information can significantly alter the revision of probability contingent upon it. Real-life problems need to be modeled before they can be solved formally. And for the selection of an appropriate model (e.g., sample space), the way in which information is obtained (information structure) is crucial.

In the case of the Harvard Medical School Test (Casscells, Schoenberger, and Graboys 1978) and in the later test by Eddy (1982), it was simply assumed that both questioner and respondent had the same model in mind. However, both were trained differently and therefore had modeled the problem differently.

5.5. A Two-Model Problem

A simple and therefore nicely illustrative example of how different models of real-world problems can lead to different rational judgments is the question of whether the "hot hand" in basketball exists. The issue is that there exist two different models of the "hot hand," one claiming that the hot hand is a "cognitive illusion" (Tversky and Gilovich 1989) and the other claiming that "it is okay to believe in it" (Larkey, Smith, and Kadane 1989).

According to Tversky and Thomas Gilovich (1989), the belief in the existence of the hot hand is caused by a misconception of randomness. It is what Tversky and Kahneman (1974) called the "gambler's fallacy" (see section 5.3). To show this they define a hot hand or "streak shooting" as a departure from coin tossing in two essential respects. Firstly, the frequency of streaks (i.e., moderate or long runs of successive hits) must exceed what is expected by a chance process with a constant hit rate. Second, the probability of a hit should be greater following a hit than following a miss, yielding a positive serial correlation between the outcomes of successive shots:

1. Number of streak hits > number of streaks by chance process.
2. $Pr(H \mid H) > Pr(H \mid M)$, where H denotes a hit and M a miss.

To investigate this, they first used the field-goal records of nine major players of 48 home games of the Philadelphia 76ers during the 1980–1981 season. These data provided no evidence for the second claim. The second study of claims 1 and 2 were controlled experiments. Twenty-six members of the varsity teams at Cornell University were recruited to participate in shooting experiments, again with no evidence for claims 1 and 2.

Patrick D. Larkey, Richard A. Smith, and Joseph B. Kadane propose a different conception of how observers' beliefs in streak shooting are based on National Basketball Association (NBA) player shooting performances. They tested this alternative conception on data from the 1987–1988 NBA season, and came to the conclusion that the hot hand does exist. They found this different result, because, as they saw it, there is a problem with Tversky and Gilovich's conceptualization as well as their data analyses of the origination, maintenance, and validity of beliefs about the "streakiness" of particular players.

According to Larkey, Smith, and Kadane, the shooting data that Tversky and Gilovich analyze are in a very different form than the data usually available to observers of streak shooting. The data analyzed by Tversky and Gilovich consist of isolated individual player shooting sequences by game, but the data "available to observers including fans, players, and coaches for analysis are individual players' shooting efforts in the very complicated context of an actual game" (p. 24). The hot-hand phenomenon is a pattern that only exists in the unfolding sequence of shot opportunities in a game rather than in the activities of an individual player, and so requires, according to Larkey, Smith and Kadane, a very different model of player shooting activities than the one used by Tversky and Gilovich. Tversky and Gilovich's model ignores the game context, how a player's shooting activities interact with the activities of the other players. As a result Larkey, Smith, and Kadane come up with a different hypothesis to be tested than the two claims for which Tversky and Gilovich found no evidence:

> The field goal shooting patterns of players with reputations for streakiness will differ from the patterns of reputationless players; a streak shooter will accomplish low-probability, highly noticeable and memorable events with greater frequency than reputationless players in the data set and with greater frequency than would be expected of him in the context of a game. (Larkey, Smith and Kadane 1989, p. 26)

In order to look not only at isolated player shooting sequences, they argued for taking the context of a game into account. They hypothesized that this context is what really enables observers of NBA basketball to differentiate streak shooters from the other players. Therefore context is defined as a sequence of 20 consecutive field goal attempts taken by all players in a game. In a game g in which all players attempted A_g field goals, the total number of 20 shot contextual sequences is $A_g - 20 + 1$. Performance of a player is then expressed as a ratio of which the numerator is the number of times that a player accomplishes the sequence of a given length in context. The denominator is an expectation: the number of shot opportunities times the probability of a player taking T or more shots (where T is greater than or equal to the sequence length, L) in a 20-field-goal context and of making j of L shot regardless of position in the T shots:

$$P_i^L \left\{ \sum_{j=L}^{20} \binom{20}{j} \gamma_i^j (1 - \gamma_i)^{20-j} (L + 20 - j) \right\} S,$$

$$S = \sum_{g=1}^{G} (A_g - 20 + 1),$$

where

P_i = Probability of a hit given a shot by player I,
L = Length of run,
γ_i = Probability of player i taking a shot,
G = Number of games,
A_g = All field goal attempts in game g.

A model of a hot hand in a game context is clearly more complex than a model of a hot hand of an isolated player.

When using this model for analyzing the data, Larkey, Smith, and Kadane found evidence for the existence of hot hands. In their conclusion they emphasized the difference between a real-life setting and an experimental setting:

> Attributing error in reasoning about chance processes requires at the outset that you know the correct model for observations about which subjects are reasoning. Before you can identify errors in reasoning and explain those errors as the product of a particular style of erroneous reasoning, you must first know the correct reasoning. It is much easier to know the correct model in an experimental setting than in a natural setting. In the experimental setting you can choose it. In a natural setting such as professional basketball you must first discover it. (Larkey, Smith and Kadane 1989, p. 30)

5.6. Monty Hall

> On a TV show, a new car is hidden behind one of three screens; the MC has the contestant select one screen. He then opens one of the two remaining screens, revealing no car, and asks the contestant whether he wishes to change his initial selection. Can the contestant increase his odds by doing so?
>
> —Barbeau 1991

The case of the existence of hot hand is a relatively simple one, because it is a choice between two models, but for real-world problems usually a larger number of models is possible. The reason for this is that a verbally stated real-life problem leaves considerable freedom in the way this problem can be translated into a probabilistic model, which implies a determination of the sample space, the target probability, and the information structure. This can be shown by the decades-long debate about the Monty Hall three doors problem, or more simply the Monty Hall problem.

According to Daniel Friedman, the Monty Hall problem is "one of the most persistent and best documented examples of irrational behavior" (Kluger and Friedman 2010, p. 31), and was therefore chosen by him as a subject for experimental research on economic decision-making (Friedman 1998)[17]:

> Host Monty Hall of the once-popular TV game show "Let's Make A Deal" asked his final guest of the day to choose one of three doors (or curtains). One door led to the "grand prize" such as a new car and the other two doors led to "zonks" or worthless prizes such as goats. After the guest chose a door, Monty always opened one of the other two doors to reveal a zonk and always offered the guest the opportunity to switch her choice to the remaining unopened door. The stylized fact is that very few guests accepted the opportunity to switch. (Friedman 1998, p. 933)

According to Friedman, nonswitching is "anomalous" because in the game just described the probability of winning is $1/3$ for nonswitchers and $2/3$ for switchers. For this outcome he referred to Stevin Selvin (1975), Barry Nalebuff (1987), Marilyn vos Savant (1990), and Leonard Gillman (1992), in which the Monty Hall problem is discussed and suggested that switching is the winning choice.

To test the three-door choice task, Friedman (1998) conducted the following experiment:

> One hundred four subjects were recruited. . . . Each subject entered a quiet room and sat at a table opposite the conductor with no other subjects present. After reading the instructions, each subject completed a series of ten trials (or "periods"). In each trial, the subject initially picked one of three

> face-down cards. Then the conductor turned over a nonprize card that the subject did not choose and offered her the opportunity to switch to the other face-down card. Finally both face-down cards were turned over, one of which was the prize card. Each trial the subject earned 40 cents when her final choice turned out to be the prize-card and 10 cents otherwise. (Friedman 1998, p. 934)

Figure 5.3 shows the results. The switch rate started out extremely low (less than 10% in the first trial) and increased fairly steadily over the next several trials. But it stagnates at about 40% after the sixth trial and actually declines in the last trial to about 30%. The overall switch percentage is 28.7%.

This result was remarkable:

> The three-door task is now a true anomaly. The data are not hypothetical: more than 100 real people left lots of real money on the table in a controlled laboratory setting. The observed behavior is not "approximately" rational; most of the people most of the time made the irrational choice when the rational choice was just as convenient. And with ten trials, each laboratory subject had the opportunity to become familiar

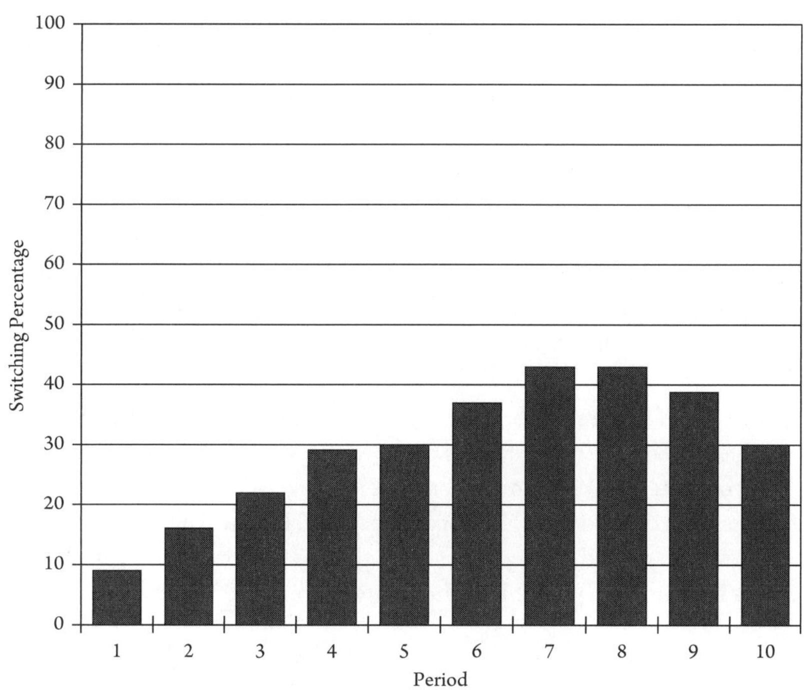

Figure 5.3 Switch Results of a Monty Hall Experiment. Source: Friedman 1998, p. 936.

> with the task and to find the rational decision. . . . One conclusion seems clear. The three-door task now deserves a place among the leading choice anomalies. . . . Indeed, I am not aware of any anomaly that has produced stronger departures from rationality in a controlled laboratory environment. (Friedman 1998, pp. 935–936)

Friedman, however, could not explain this result, "the ultimate reasons for its strength remain unclear" (p. 936). So the second part of his article dealt with the following two questions: "Why are people so irrational? When (if ever) can people become more rational in tasks of this sort?" (p. 937). Therefore, he ran the same baseline procedures, but now with additional alternative treatments ("Run 2"), chosen to encourage more effective learning: higher incentives, track record, written advice and comparison of results, and more trials (15 instead of 10).

The written advice is of particular interest for understanding the role of different modeling options. Before the first trial of "Run 2," a subject in this treatment was handed a page with two paragraphs in random order. One paragraph recommended always switching and explained succinctly why switching improves the odds. The other paragraph, written in a similar style, recommended always not switching. The paragraphs were not presented separately to avoid confounding demand effects with learning through written advice. The text was as follows:

> *Advice treatment text.*—Please read the [following] two pieces of advice on how to earn money in this experiment. The pieces disagree on what you should do; you must make up your own mind.
>
> Advice S1: You have 1 chance in 3 of picking the Prize card initially. If you did pick it initially you will win the prize if you remain. You have 2 chances in 3 of not picking the Prize card initially. If you did not pick it initially you will win the Prize if you Switch. So, the overall chance of winning is 2 chances in 3 if you Switch and only 1 in 3 chances if you Remain. Therefore, you will do the best in the long run if you always Switch.
>
> Advice R1: The conductor may try to distract you by offers to Switch, but these offers are made for his/her own reasons and are not necessarily in your interest. When only 2 cards remain face down, you have at least 1 chance in 2 of picking the Prize card and Switching will not improve your chances. You should not let yourself be distracted. Therefore, you should always Remain with your initial choice and never Switch. (Friedman 1998, pp. 942–943)

Even though the additional treatments increased the overall switch rate, the effect of these treatments was not big: it rose to a maximum of 55% in period 10 but trailed off somewhat in the last few periods. Nevertheless Friedman concluded that "every choice 'anomaly' can be greatly diminished or entirely eliminated in appropriately structured learning environments" (p. 941).

There is, however, in my view, something unsatisfactory about this experimental result. In the history of the Monty Hall problem and the discussions that went with it (more about this below), two aspects are striking: there are only two camps and the dividing line is sharp: one camp, say *NS*, believes very strongly that switching is irrelevant, and the other camp, say *S*, believes very strongly that switching is optimal. When people switch from camp *NS* to camp *S*, this is always permanent and irreversible. No one switches from *S* to *NS*.

The deeper question, however, is this: Did Friedman's experiment show that people are "so irrational"? To investigate this question, I have studied how this Monty Hall problem is conceived and discussed in mathematical and statistical journals over four decades. The reason for investigating this specific literature is that the participants in the discussions are experts in mathematical and probabilistic reasoning, and so cannot be put aside as being "naive" or "intuitive."

Important for assessing whether a judgment is rational is to figure out which model has been used to account for the problem at hand. This is often not obvious. One has to infer this in each case from the verbal explanation of the problem and the suggested solution given along with it in that specific case. In the literature one will find several different verbal expressions of the Monty Hall problem; compare, for example, the epigraph at the beginning of this section with the one given by Friedman. The current standard verbal expression only became stabilized after the publication of Marilyn vos Savant's (1990) discussion of it in her own well-read column "Ask Marilyn" in *Parade Magazine*. Because most of the more recent discussions of the Monty Hall problem refer to her wording of it, we take hers as the starting point for discussing the various proposed model options:

> Suppose you're on a game show, and you're given the choice of three doors. Behind one door is a car, behind the others, goats. You pick a door, say #1, and the host, who knows what's behind the doors, opens another door, say #3, which has a goat. He says to you, "Do you want to pick door #2?" Is it to your advantage to switch your choice of doors? (vos Savant 1990, p. 13)

To better understand what kind of model she had in mind, her answer also needs to be given (due to the fact that the choice decision and model are inextricably linked with each other, the proposed answer helps to uncover the kind of model she has used): "Yes; you should switch. The first door has a $\frac{1}{3}$ chance of winning, but the second door has a $\frac{2}{3}$ chance" (vos Savant 1990, p. 13). She, subsequently, gave four explanations why this answer should be considered as the only correct one.

Before the model will be inferred from vos Savant's problem setting and her solution to it, let us first define a few symbols that will be used in the rest of this chapter. Let C_i denote the event that the car is at door i, and H_j the event that the host opens door j $(i, j = 1, 2, 3)$. Without any loss of generality we can say that the contestant chooses door 1. W_s denotes the event that you win the car if you switch. Let us define the following probabilities: $\pi_i = Pr(C_i)$ and $p_{ij} = Pr(H_j \mid C_i)$. Note that the probabilities p_{ij} represent the host's possible strategies. The selection of the appropriate events, C_i and H_j $(i, j = 1, 2, 3)$, determines the sample space of the model. The restrictions put on the host's strategy, p_{ij}, determines the information structure of the model. And the definition of the probability of winning a car is the target probability of the model.

Vos Savant's solution can now be modeled as follows. Assume that $\pi_i = 1/3$ and let us only assume the host will always open a door: $p_{i2} + p_{i3} = 1$. The simplest proof of the vos Savant's answer is the following:

$$Pr(W_s) = Pr(H_2 \text{ and } C_3) + Pr(H_3 \text{ and } C_2) = \pi_3 p_{32} + \pi_2 p_{23} = 1/3 1 + 1/3 1 = 2/3. \tag{5.1}$$

A more general result than vos Savant's can be obtained by permitting that the host might also open a door with a car behind it:

$$\begin{aligned} Pr(W_s) = {} & Pr(H_2 \text{ and } C_2) + Pr(H_2 \text{ and } C_3) + Pr(H_3 \text{ and } C_2) \\ & + Pr(H_3 \text{ and } C_3) = 1/3(p_{22} + p_{32} + p_{23} + p_{33}) = 2/3. \end{aligned} \tag{5.2}$$

It should be noted that these results are independent of the strategy of the host. The same result would be obtained when the host opens only a door he knows will not show a car: $p_{22} = p_{33} = 0$, and thus $p_{23} = p_{32} = 1$; or when he opens a door randomly $p_{22} = p_{23} = p_{32} = p_{33} = 1/2$.

Vos Savant's solution initiated a lively debate among mathematicians and statisticians about whether she was right or wrong. It is this debate in particular that shows that there is not simply one unique model of the Monty Hall problem, and thus one unique way to make a rational judgment when faced with this problem. This debate was actually a debate about what the "correct" interpretation of the problem is in terms of the sample space, the target probability, and the information structure.

One of the first publications that emphasized that it matters which target probability is chosen was V. V. Bapeswara Rao and M. Bhaskara Rao (1992). They consider equation (5.1) as the solution to a so-called scenario 1: "The participant contacts a statistician as to the best course of action to be taken to maximize the probability of winning the automobile before getting on to the stage" (p. 90). "Scenario 2 treats the situation in which "the participant is actually on the stage" (p. 91). For this latter scenario a *conditional* probability is worked out:

$$Pr(C_1 | G_3) = Pr(C_1 | C_2 \text{ or } C_1) = Pr(C_1)/[Pr(C_2) + Pr(C_1)] = 1/3/2/3 = 1/2, \tag{5.3}$$

where G_i denotes that a goat is shown behind door *i*. According to Bapeswara Rao and Bhaskara Rao, however, this conditional probability is not the correct interpretation of vos Savant's problem. They assumed that this interpretation was the one of those who criticized vos Savant's solution.

As a matter of fact, the various verbal descriptions of the Monty Hall problem, including the two given here, do not say clearly enough what kind of target probability is assumed. The instructions of Friedman's experiment are also not explicit about the target probability, so a subject may well choose advise S1 when he or she assumes that $Pr(W_s)$ is the target (eq. 5.1 or eq. 5.2) or advise R1 when he or she assumes that the conditional probability of equation 5.3 is the target probability.

Another important difference between scenario 1 and scenario 2 is whether the choice behavior of the host is taken into account (in eq. 5.3 H_j does not appear), or in other words whether the host (H) is part of the sample space. This difference was also noted by Donald Granberg and Thad A. Brown (1995, p. 717): "Most people seem to ignore, or at least do not adequately take into account, the knowledge of the host as a cue." The sample space of Bapeswara Rao and Bhaskara Rao's model of scenario 2 does not include events H_j, but only C_i. In other words, they interpret the problem as having the sample space $\{C_1, C_2, C_3\}$. Most people seem to solve the following problem: What is the (conditional) probability of winning a car by switching doors when door 3 shows a goat? Any information about procedures to open doors or the specific role of the host is not taken into account.

Vos Savant (1991, p. 347), however, emphasized that the host's strategy is irrelevant for the target probability: "Pure probability is the paradigm, and we published no significant reason to view the host as anything more than an agent of chance who always opens a losing door and offers the contestant the opportunity to switch." So, in vos Savant's model, the host is in the sample space, but the host's strategy does not influence the target probability. See for this also the comment just below equation 5.2. Nevertheless, in the mathematical and statistical literature the idea took hold that the strategy of the host does actually matter. The publication that contributed particularly to this idea about the relevance of the host's strategy is J. P. Morgan, N. R. Chaganty, R. C. Dahiya, and M.J. Doviak (1991a).

According to Morgan and coauthors (1991a), vos Savant was right about the answer to switch, but for the wrong reasons. Therefore, they discussed six, in their view, false solutions. To see that these solutions were considered to be false, one should notice that the authors had changed the original vos Savant problem a tiny bit, but enough to make a difference:

> Suppose you're on a game show and given a choice of three doors. Behind one is a car; behind the others are goats. You pick door No. 1, and the host, who knows what's behind them, opens No. 3, which has a goat. He then asks if you want to pick No. 2. Should you switch? (Morgan et al. 1991a, p. 284)

Or in their own words:

> The player has chosen door 1, the host has then revealed a goat behind door 3, and the player is now offered the option to switch. (Morgan et al. 1991a, p. 284)

The crucial difference is that the host does not open *any* of the other doors with a goat behind it, which is indicated by using the expression "say" in vos Savant's formulation of the problem, but a specific door, namely door No. 3. In other words, contrary to Bapeswara Rao and Bhaskara Rao (1992), Morgan et al. (1991a) consider the conditional probability as the correct target probability of vos Savant's problem. This can more clearly be seen from their discussions of the six "false solutions," in particular the first one:

> *Solution F1*. If, regardless of the host's action, the player's strategy is to never switch, she will obviously win the car $1/3$ of the time. Hence the probability that she wins if she does switch is $2/3$. (Morgan et al. 1991a, p. 284)

In their comments on this solution, they point explicitly to the conditional probability as the, in their view, correct target probability: "F1 is a solution to the unconditional problem. . . . The distinction between the conditional and unconditional situations here seems to confound many, from whence much of the pedagogic and entertainment value is derived" (p. 285). Three other solutions (F2, F3, and F5) are also considered to be "true" solutions but again to the unconditional problem, which is not, according to Morgan and coauthors (1991a), the correct interpretation of the problem.

Their solution F4 gives an answer to the conditional problem, but now the original sample space is incorrectly specified, with the result that the probability of winning by switching is $1/2$.

> *Solution F4*. The original sample space and probabilities are as given in Solution F2. However, since door 3 has been shown to contain a goat, GGA is no longer possible. The remaining two outcomes form the conditional sample space, each having probability $(1/3)/(1 - (1/3)) = 1/2$. (Morgan et al. 1991a, p. 285)

If one defines the sample space as $\{C_1, C_2\}$, the chance of winning a car is indeed $1/2$.

Solution F6 gives, according to the authors, a correct specification but assumes falsely a certain strategy on the part of the host. They showed that a more general solution can be given. The general solution they give for the conditional problem is

$$Pr(W_s \mid G_3) = Pr(H_3 \text{ and } C_2) / [Pr(H_3 \text{ and } C_1) + Pr(H_3 \text{ and } C_2)] \\ = p_{23} / (p_{13} + p_{23}). \tag{5.4}$$

The problem is then solved by the assignment of values for the p_{ij}s, viewed as a quantification of the host's strategy. For example, in solution F6 the quantification $p_{23} = 1$ and $p_{13} = 1/2$ was assumed.

Because the host never has the option of showing the car, which they call the "vos Savant scenario," the host's strategy can be specified as follows:

$$p_{12} = p, p_{13} = q = 1 - p, p_{22} = p_{33} = 0, p_{23} = p_{32} = 1. \tag{5.5}$$

So

$$Pr(W_s \mid G_3) = 1/(1 + q) \tag{5.6}$$

and thus they arrive at the solution that $Pr(W_s \mid G_3) \geq 1/2$ for every q. In other words, when we have a vos Savant scenario, "we need not know, or make any assumption about, the host's strategy to state that the answer to the original question is yes. The player should switch, for she can do no worse and may well improve her chances" (p. 286).

To sum up, Morgan and coauthors (1991a) discussed six "scenarios," which differ with respect to assumptions about the target probability, sample space, and information structure. For each scenario these assumptions determine a specific "solution." Although for each scenario the inferred solution is correct, the solutions were nevertheless considered to be "false" because each of them was considered to be based on "false" assumptions. But what was considered to be a false assumption depended on their interpretation of what would be the "true" assumptions of the Monty Hall problem.

In the last part of their article, the authors also give their solution to the unconditional problem, "for it evaluates the proportion of winners out of all games with the player following a switch strategy" (p. 286):

$$Pr(W_s) = Pr(W_s \mid G_2)Pr(G_2) + Pr(W_s \mid G_3)Pr(G_3) = \\ \frac{p_{32}}{p_{12} + p_{32}} \frac{p_{12} + p_{32}}{3} + \frac{p_{23}}{p_{13} + p_{23}} \frac{p_{13} + p_{23}}{3} = \frac{p_{23} + p_{32}}{3}.$$

So, according to this solution, one cannot do better than $2/3$ in the unconditional game, and the vos Savant scenario "maximizes the overall efficacy of the switch strategy" (p. 286).

If Morgan and coauthors (1991a) are right with respect to this latter solution, then the unconditional probabilities depend also on the host's strategy, which would contradict the solutions of the unconditional probabilities discussed above.

But they are not right. The first part of the above solution shows that the assumed sample space is $\{G_2, G_3\} = \{(H_2 \text{ and } C_1), (H_2 \text{ and } C_3), (H_3 \text{ and } C_1), (H_3 \text{ and } C_2)\}$. This sample space, implicitly, enforces the so-called "vos Savant scenario" (see eq. 5.5); thus $p_{23} = p_{32} = 1$. Therefore, the unconditional probability is $Pr(W_s) = 2/3$. So even these authors who aimed at a "wider dissemination for instructional purposes" of the "pitfalls of conditional probability calculations and interpretations" (p. 284) could not help adding another "pitfall" to the literature. This specific "pitfall" arises whenever one does not make a distinction between the sample space and the host's strategy.[18]

In a reply, vos Savant (1991) complains about "strong attempts at misinterpretation" of the original question, and therefore restates her original interpretation once again: the target probability is the unconditional probability. "Nearly all of my critics understood the intended scenario, and few raised questions of ambiguity" (vos Savant 1991, p. 347). In a rejoinder in the same rubric, however, Morgan and coauthors (1991b) maintain the claim that the problem should be considered a conditional probability problem.

The Monty Hall debate initiated by vos Savant's discussion of this choice problem shows that it is important to distinguish between the target probability, the sample space, and the information structure, and that all three should be made explicit; otherwise, the discussion about the correct model will be unnecessarily confused by additional flaws. The most important point that the discussion in the mathematical statistical literature shows is that there is not one "true" way to model the Monty Hall problem. In all three dimensions of target probability, sample space, and information structure, modeling decisions have to be taken that depend on specific interpretations of what the Monty Hall problem entails.

Various authors I have discussed considered the Monty Hall paradox as exemplary for the kind of difficulty one has to deal with when modeling real-world problems in contrast with textbook problems: "Textbook problems are often set in artificial situations that neither require nor inspire real-world thinking on the student's part" (Morgan et al. 1991a, p. 284); see also the remark by Bar-Hillel and Falk in section 5.4. Modeling a real-world problem presupposes drawing up assumptions that formalize the problem as appropriately as possible. Modeling of a real-world problem is not an obvious task, and one can have different interpretations of what the problem entails. The Monty Hall problem shows that beside the sample space and information structure, the target probability should also be specified.

Alan H. Bohl, Matthew J. Liberatore, and Robert L. Nydick (1995) discuss the importance of assumptions in problem solutions.

> We believe that the issue of clearly stating, then questioning, and finally agreeing on, one or more sets of appropriate assumptions is critical to the success of any modeling effort. "Buying in" to the assumptions is

tantamount to buying in to the solution. In the long run, you cannot have the second without the first. (Bohl, Liberatore, and Nydick 1995, p. 4)

They give the following "reasonable" set of assumptions, which they consider as consistent with the solution that switching increases the probability of winning to $^2/_3$.

1. The game show host will always open a door that was not selected by the contestant.
2. The opened door will always reveal a goat.
3. If the contestant selects a car, the game show host will select one of the other two doors with equal probability.
4. The car is more valuable than the goats.
5. The position of the car will not change once the game begins.
6. The contestant cannot offer to sell the door to the host or to anyone else.
7. The game show does not have the option to offer the contestant money to purchase his or her door.
8. The game show host will always offer the opportunity to switch.
9. The game is not repeatable, that is, the contestant only has the opportunity to play the game once.
10. The initial location of the car was randomly determined prior to the start of the game. (Bohl, Liberatore and Nydick 1995, pp. 4–5)

These assumptions, however, show that the three model dimensions have already been specified. These model specifications, that is, target probability, sample space, and information structure tell which additional assumptions (like those listed here by Bohl, Liberatore and Nydick 1995) are needed. Another model specification will require other assumptions of further specification. As has been shown in this chapter, modeling a real-world problem entails first deciding the target probability, the sample space, and the way information is provided. In all three directions, assumptions about their properties have to be made explicit before arriving at a definite solution.

So, confronted for the first time with a verbally expressed real-life decision problem, it is often not clear what the target probability, sample space, or information structure is, unless stated explicitly. Different interpretations of these three dimensions lead to different rational outcomes. As such, a debate about which decision is most rational should be a debate about which model is considered to be the most appropriate representation of the problem and its settings.

This conclusion has clear implications for running experiments on probabilistic decision making. Friedman (1998, p. 941) asserts that "Every choice 'anomaly' can be greatly diminished or entirely eliminated in appropriately structured learning environments." People are "not hardwired to behave irrationally" (p. 941), but the task may not be easy to learn. According to him, "there is no such thing as an anomaly in the traditional sense of stable behavior that is inconsistent with rationality. There are only pseudo-anomalies describing transient behavior before the learning process has been completed" (p. 941). As a consequence, absence of adequate learning opportunities leads to "pseudo-anomalies."

I can only agree with this assertion, but have a somewhat different view on what this learning entails. To a subject in an experiment, it is often not clear what the experiment's model is. Note that even the instructions usually given at the beginning of an experiment do not reveal them; see, for example, Friedman's "Advice Treatment Test" above. This advice, for example, does not tell what the target probability is. So the subject will have his or her own interpretation of the task, which may deviate from the experimenter's design. As a result, the subject will make decisions that within the experimental setup look irrational. But if the experimenter explains what the model assumptions are, for example, by explicitly telling the subject what the target probability is, the subject will make the same "rational decision" as expected by the experimenter, that is, there will be no anomaly. As you can see from the mathematics used above, the mathematical derivations are not complicated, so no "special training" is really needed.

Another way of learning is to explain the problem by playing the game as a specific card game, such as in Friedman's experiment. Each card game is of course a specific interpretation of the choice problem. But as with any other card game, by playing it several times we can learn the winning strategies. So by playing the Monty Hall game with cards enough times, it will give the subject the opportunity to grasp in a more intuitive way the model's assumptions (learning by doing, or perhaps it is better to say, learning by playing). Then, like all of us, in the end the subject will learn what the most rational choice is for this specific task. Some people take more time to see the model than others, so the amount of trials should be sufficient for an average subject to learn. It even may be the case that in Friedman's experiment, the increase of the switch rate in Run 2 was simply caused by the increase of the number of trials.

5.7. Conclusions

We now can be more precise about what kind of judgment is required for measurement outside the laboratory, by first saying what it is not. It is not "rational judgment." A rational judgment presupposes a model of which it is the optimal outcome. From a Kantian perspective, one could doubt whether it is a judgment. Judgment is "considered judgment" that has to include considerations about what the most appropriate model would be, that is, considerations about target probability, information structure, and sample space. Considered judgment precedes rational judgment. To say that subjective judgment is "biased," or "irrational," has no meaning as long as one does not specify the model that defines when a decision is irrational or a measurement is biased.

Judgment required for measurement outside the laboratory is "scientific judgment"; it includes expertise about the phenomena being studied and measured, "of actually knowing something about real phenomena, and of making realistic assumptions about them" (Haavelmo 1944, p. 29), and knowledge about the required tools and techniques. But it also includes scholarly knowledge, that is, knowing what is already known about the phenomenon—mind the "theoretical" link in the measurement chain of figure 5.2 that is represented by a human figure sitting at a table reading the material from the "literature store." But on top of this, judgment also refers to imagination, imagining features of the phenomenon that have not been observed yet, but which you imagine Nature may not exclude.

The kind of subjective knowledge that is needed to complement objective knowledge is knowledge that is not crystallized in models and is personal. It is knowledge that is part of Karl Popper's "World 2": "the mental or psychological world, the world of our feelings of pain and of pleasure, of our thoughts, of our decisions, of our perceptions and our observations; in other words, the world of mental or psychological states or processes or of subjective experiences" (Popper 1979, p. 1). Models belong to Popper's "World 3": the world of objective knowledge is "the world of the products of the human mind, such as languages; tales and stories and religious myths; scientific conjectures or theories, and mathematical constructions; songs and symphonies; paintings and sculptures" (p. 2).

Although I argue that subjective judgments cannot be validated in terms of being "rational" or "unbiased," they can be empirically validated in ways comparable to those that validate the products of World 3. Methods have been developed to assess the possessor of personal knowledge, that is, the expert. These methods will be discussed in the next chapter.

Appendix: Statistical Bias

In the literature discussed in this chapter, it is assumed that Bayesian reasoning is an "unbiased" heuristic. In mathematical statistics, however, unbiasedness has a very specific meaning: An estimator $\hat{\theta}$ is unbiased if and only if its expected value is equal to the parametric value, θ, it is intended to estimate: $E[\hat{\theta}] = \theta$. A consequence of this statistical definition is that an estimator based on Bayesian reasoning is not automatically unbiased. For example, in a widely used standard textbook on statistics, *Introduction to the Theory of Statistics* (Mood, Graybill, and Boes 1974), one will find the following notable observation: "in general a posterior Bayes estimator is not unbiased" (p. 343).[19] So a freshman in statistics is warned that Bayesian tools and unbiasedness might be incompatible.

Being warned, let us check whether the post-test probability, that is, the probability taking account of test results, $Pr(P \mid X)$, is an unbiased estimator of the pretest probability, $Pr(P)$. Let X be the random variable indicating the test result, taking value + or –; then

$$E\left[Pr\left(P|X\right)\right] = E\left[\frac{Pr\left(X|P\right)}{Pr\left(X\right)}\right] Pr\left(P\right)$$
$$= \left[\frac{Pr\left(+|P\right)}{Pr\left(+\right)} Pr\left(+\right) + \frac{Pr\left(-|P\right)}{Pr\left(-\right)} Pr\left(-\right)\right] Pr\left(P\right) = Pr\left(P\right).$$

So it seems that our worry was unnecessary. Unfortunately, this is not the case. Generally, in fields (including evidence-based medicine) where clinical decision-making is meant to be rational, it is highly recommended one use likelihood ratios to estimate the disease odds. When discussing the use of likelihood ratios, Roger M. Cooke (1991) gives an expression of how one can "learn" from observations (adapted from his theorem 6.3, p. 97):

$$E[LR(X) \mid P] \geq 1, \text{ and equality holds if and only if } Pr(LR(X) = 1 \mid P) = 1.$$

The equality condition can hold only if $Pr(X \mid P) = Pr(X \mid A) = Pr(X)$. A test that would have this latter characteristic is not informative because it is then independent of disease, and should therefore be excluded; see section 5.2.

However, this theorem shows only that one can learn from a test in case the disease is present. The surprising result, however, is that in case of an absent disease, a test will not "teach" us about the absence of this disease:

$$E[LR\left(X\right) \mid A] = \frac{Pr\left(+|P\right)}{Pr\left(+|A\right)} Pr\left(+|A\right) + \frac{Pr\left(-|P\right)}{Pr\left(-|A\right)} Pr\left(-|A\right) = 1.$$

This remarkable result makes the (Bayesian) assessment of a test result biased:

$$E\left[LR(X)\right] = E[LR(X) \mid P]Pr(P) + E[LR(X) \mid A]Pr(A) > 1.$$

So it appears to be the case that post-test odds are not unbiased estimators for the pretest odds:

$$E\left[\frac{Pr\left(P|X\right)}{Pr\left(A|X\right)}\right] = E\left[LR\left(X\right)\right] \cdot \frac{Pr\left(P\right)}{Pr\left(A\right)} > \frac{Pr\left(P\right)}{Pr\left(A\right)}.$$

The undesired result of this bias is that each time a test result is accounted for (whatever the result is, positive or negative) the expected disease odds will increase.

Notes

1. The Evidence-Based Working Group was chaired by Gordon Guyatt.
2. See also Karthikeyan and Pais (2010), for similar arguments.
3. This brief account on judgment is based on Fleischacker (1999). I thank Thomas Wells by bringing Fleischacker (1999) to my attention.
4. Quoted in Buddingh (1989). Henk Don was director of CPB Netherlands Bureau for Economic Policy Analysis from 1994 to 2006 (on which more in chapter 6).
5. *Fiddler on the Roof* is a 1971 musical film with screenplay written by Joseph Stein.
6. It is no accident that as well in the literature on measurement as on rational decision-making the same authors appear, most prominently Amos Tversky. See Heukelom 2010 for an excellent history about this close connection.
7. See also the discussion of the "rational consensus approach" in chapter 6, according to which experts are being treated as measuring instruments that need to be calibrated to arrive at a "rational consensus."
8. Tversky and Kahneman (1974, p. 1124) compared the assessment of probability with the "assessment of physical quantities such as distance or size," so they saw quite explicitly a close relationship between rational judgment and measurement.
9. Etymologically a paradox has the following meanings: a "statement contrary to common belief or expectation," a "statement seemingly absurd yet really true," "contrary to expectation, [something] incredible," a "statement that is seemingly self-contradictory yet not illogical or obviously untrue" (Harper 2001–2014).
10. $Pr\left(P|+\right) = \frac{Pr\left(+|P\right)Pr(P)}{Pr\left(+|P\right)Pr(P)+Pr\left(+|A\right)Pr(A)} \approx \frac{0.001}{0.001+0.05\cdot 0.999} = 0.02$, where P: disease is present, A: disease is absent, and $+$: positive test result.
11. $Pr\left(ca|+\right) = \frac{Pr\left(+|ca\right)Pr(ca)}{Pr\left(+|ca\right)Pr(ca)+Pr\left(+|benign\right)Pr(benign)} = \frac{0.792\cdot 0.01}{0.792\cdot 0.01+0.096\cdot 0.99} = 7.7\%$.
12. If $Pr(P) < T_{trx}$ and $Pr\left(P|+\right) = \frac{Pr\left(+|P\right)}{Pr(+)}Pr\left(P\right) > T_{trx} \rightarrow \frac{Pr\left(+|P\right)}{Pr(+)} > \frac{T_{trx}}{Pr(P)} > 1$.
13. This can be seen by maximizing $Pr(P|+)Pr(A|-)$ for $Pr(P)$.
14. When $Pr(+|P) \approx Pr(-\mid A)$, then also $Pr(-|P) \approx Pr(+\mid A)$, and thus $LR(+)LR(-) \approx 1$.
15. $Pr(+) = Pr(+\mid P)Pr(P) + Pr(+\mid A)Pr(A) \approx Pr(+\mid P)0.5 + Pr(-\mid P)0.5 = 0.5$.
16. Such an instruction can be found in, for example, an explicit account of this training, Sackett et. al. (2000), *Evidence-Based Medicine: How to Practice and Teach EBM*.
17. Both publication dates seem to reveal Friedman's enduring interest in the Monty Hall problem, an impression that is confirmed by a website on this problem, maintained by Aadil Nakhoda under the supervision by Friedman: "The Learning and Experimental Economics Projects of Santa Cruz! Monty Hall Problem" (LEEPS) website <http://leeps.ucsc.edu/misc/page/monty-hall-puzzle/>.
18. This is not the only "pitfall" I found in the literature, which unfortunately complicates the review of the discussions in this literature even more. See Boumans 2011 for a discussion of these "pitfalls."
19. A "posterior Bayes estimator" is defined as $E[Y \mid X]$, where X is a random variable with probability $Pr(X|Y = y)$, and Y a random variable with probability $Pr(Y)$. A posterior Bayes estimator is an "unbiased" estimator of y when $E[E[Y \mid X] | y] = y$. It is shown that a posterior Bayes estimator is unbiased only when this estimator correctly estimates y with probability one. In all other cases the estimator is not unbiased.

6

Consensus

Science aims at rational consensus, and methodology of science must serve to further this aim.

—Roger Cooke, *Experts in Uncertainty* (1991).

If we, economists, continue to oppose each other, we fail in our duty as scientists.

—Jan Tinbergen, "The Need of a Synthesis" (1982).

6.1. Introduction

The two most dominant ideals of scientific knowledge are that it is "objective" and "unified." Although the specific meanings of both terms can vary across different scientific contexts, they share the following connotations: "Objective" denotes knowledge that is impersonal, intersubjective, unbiased, and disinterested, and "unified" implies a consensus and sharing of language, laws, method, and facts. But the ideals of science as objective and unified are—unsurprisingly—not so easy to attain in scientific practice. The problems of attaining objectivity in the context of a field science have been discussed in the preceding chapters, particularly chapter 3. Consensus is an end-state of a process for which hardly any methodology exists. This chapter discuss several options for such a methodology.[1]

Perhaps the circumstances that come closest to the ideal of unification is the situation described by Thomas Kuhn's concept of "paradigm," which he uses to characterize "normal science." In the preface to *The Structure of Scientific Revolutions*, Kuhn (1970) clarifies how he arrived at his particular interpretation of the term "paradigm." He had spent the year 1958–1959 at the Center for Advanced Studies in the Behavioral Sciences at Stanford University and observed the widespread disagreement that according to him distinguished the social from the natural sciences:

> Spending the year in a community composed predominantly of social scientists confronted me with unanticipated problems about the differences between such communities and those of the natural scientists among

> whom I had been trained. Particularly, I was struck by the number and extent of the overt disagreements between social scientists about the nature of legitimate scientific problems and methods. Both history and acquaintance made me doubt that practitioners of the natural sciences possess firmer or more permanent answers to such questions than their colleagues in social science. Yet, somehow, the practice of astronomy, physics, chemistry, or biology normally fails to evoke the controversies over fundamentals that today often seem endemic among, say, psychologists or sociologists. Attempting to discover the source of that difference led me recognize the role in scientific research of what I have since called "paradigms." These I take to be universally recognized scientific achievements that for a time provide model problems and solutions to community of practitioners. (Kuhn 1970, pp. vii–viii)

Kuhn interpreted the existence of extensive disagreement within one discipline as a clear indication that the specific discipline is not yet a science.

One of the major problems for a social field science is that the social world is too complex, too idiosyncratic, too messy, and too irregular to be captured sufficiently in our models, theories, or laboratories (Bogen and Woodward 1988). This problem is discussed in chapter 4, "The Problem of Passive Observation," but that chapter has a rather open ending with respect to how to arrive at completeness. Chapters 1 and 3 show that closure is attained by adding knowledge about idiosyncratic conditions, that is, by adding expert judgment.

Experts are the carriers of expert knowledge. But expert knowledge is personal, and experts tend to disagree. This chapter will discuss methodologies of consensus, consensus not only among the involved experts but also about the model of the measurand.

Unification, however, is merely a relation between same items, like a relation between concepts or terms, or it is a relation between theories, or between methods, and so on. But in this chapter we will discuss the relation between dissimilar items, like models, statistics, and expert judgments. The integration of dissimilar sources of knowledge is—in social science—called "triangulation."

Triangulation is the strategy of using more than one method to validate a judgment on the grounds that we can have more confidence in a certain result when different methods are found to "converge," that is, be congruent and yield comparable data (Jick 1979). The triangulation metaphor is from navigation, which uses multiple reference points to locate an object's exact position. "Given basic principles of geometry, multiple viewpoints allow for greater accuracy." Similarly, "researchers can improve the accuracy of their judgments by collecting different kinds of data bearing on the same phenomenon" (Jick 1979, p. 602).

Triangulation is comparable to "reproducibility" (see also chapter 2), but is—instead of reproducibility—more often used in social science. Triangulation shares with reproducibility the same basic assumption that the weakness in each single method will be compensated by the counterbalancing strengths of another. That is, it is assumed that multiple and independent measures do not share the same weakness or potential for bias. "Although it has always been observed that each method has assets and liabilities, triangulation purports to exploit the assets and neutralize, rather than compound, the liabilities" (Jick 1979, p. 604). When there is convergence, confidence that the results are not attributable to a method's artifact has increased.

Todd D. Jick (1979), who discusses triangulation extensively, notes that triangulation is useful even without convergence. When divergent results emerge, "the researcher may uncover unexpected results or unseen contextual factors" (p. 608).[2]

> Triangulation may also help to uncover the deviant or off-quadrant dimension of a phenomenon. Different viewpoints are likely to produce some elements which do not fit a theory of model. Thus, old theories are refashioned or new theories developed. Moreover . . . divergent results from multimethods can lead to an enriched explanation of the research problem. (Jick 1979, p. 609)

To decide whether or not results have converged requires, however, "considered judgment" (see chapter 5). In practice, there are no guidelines for systematically ordering eclectic data in order to determine congruence or not. For example, should all components of a multimethod approach be weighted equally, that is, is all the evidence equally useful? If not, then it is not clear on what basis the data should be weighted, aside from subjective judgment.

> One begins to view the researcher as builder and creator, piecing together many pieces of a complex puzzle into a coherent whole. It is in this respect that the first-hand knowledge drawn from qualitative methods can become critical. While one can rely on certain scientific conventions (e.g. scaling, control groups, etc.) for maximizing the credibility of one's findings, the researcher using triangulation is likely to rely still more on a "feel" of the situation. This intuition and first-hand knowledge drawn from the multiple vantage points is centrally reflected in the interpretation process. (Jick 1979, p. 608)

Expert judgments are unavoidably "personal" and "subjective" and therefore any triangulation methodology has to include a strategy to ensure objectivity. Strategies that have been developed for this purpose share the idea that individual expert judgments have to be combined such to reduce subjectivity and bias.

Two methods of combining expert judgments are used in economics, the committee-consensus method and the Delphi method. In chapter 2, I have argued that arriving at a more objective Type B uncertainty evaluation requires that it be model based. Therefore this chapter will also discuss cases of expert consensus, where the combination of experts is instantiated by the measurement model.

6.2. Committee-Consensus Method

A good example of the committee-consensus method is its application in the process of decision-making at the Bank of England, where decisions have to be made on setting the interest rate and other monetary policy instruments.

> A number of central banks around the world use some form of committee structure when managing monetary policy, and there are good reasons for this. Placing total control for setting interest rates in the hands of a single unelected official would seem like a very risky proposition to most politicians. And there is evidence to suggest that groups of experts make better decisions than individuals when dealing with technical issues such as monetary policy. They can share information and learn from each other, and they can change their minds in the face of sound arguments. (Lambert 2005, p. 56)

At the Bank of England, these decisions are made by the Monetary Policy Committee (MPC). According to Paul Downward and Andrew Mearman (2008), who have investigated this process and suggest calling it "triangulation," decision-making by the MPC "appears to reflect a more pragmatic and pluralist approach that draws upon a variety of sources of argument and evidence" (p. 385).

The Monetary Policy Committee comprises the governor, two deputy governors, and six other members. Two of these six members are internal members taking management responsibilities for monetary policy and market operations. The remaining four members are "recognized experts" (Rodgers 1997, p. 244). These external members are typically academic or professional economists. Robert Lambert (2005), a former member of the Monetary Policy Committee, provided a useful inside view on the dynamics of the committee. According to him, the argument for including external members is that "they bring in a wider range of expertise and experience than would be available if the MPC could draw only on the Bank's own staff. And they bring fresh thinking to the Committee since they are only there for a limited period" (p. 56). They are not there to represent any particular interest. They have been chosen for their particular expertise. "And their responsibility is to the country as a whole, rather than to any sectional interest" (p. 56).

The decisions are made by a vote of the Monetary Policy Committee, with each member having one vote. "The crucial point is that all nine members of the Committee are individually held to account for their decisions" (Lambert

2005, p. 56). Their separate votes are recorded and published. The committee is an "individualistic committee" and is not required to reach a consensual decision but a conclusion of the majority (Downward and Mearman 2008, p. 391). Nevertheless, at the meeting there is room for discussion and expressing views. The governor asks

> the Deputy Governor responsible for monetary policy, to speak first and to give her view. She will talk for roughly ten minutes, highlighting the issues that she has thought most relevant in the previous weeks and explaining the thinking that has led her to make her decision on the rate, which she announces. The Governor then goes at random around the table and asks each member to give a similar presentation of his or her views followed by their decision on the rate, again lasting for about ten minutes or so, and usually working from prepared notes. Occasionally members will say that they would prefer to hear other people's views before they cast their vote, and so will hold their final judgement until the end of the discussion. After each person has spoken, the Governor invites questions: he himself speaks, and votes, last. (Lambert 2005, p. 59)

So, the Monetary Policy Committee method of reaching consensus among the experts is by voting. Models play only a modest role, as one of the information sources. They do not dominate the process of reaching consensus. A method in which models play a more dominant role in bringing about consensus om the decisions to be made is the method used in the US Troika process (Donihue and Kitchen 2000).[3]

The Troika is an institutional structure for generating economic forecasts and for evaluating and coordinating macroeconomic policy within the executive branch of the US government and is made up of representatives from the three groups with primary responsibility for economic and budget issues: The Council of Economic Advisers (CEA), the Office of Management and Budget (OMB), and the Department of the Treasury. Membership in the Troika is divided across three levels, with political appointees in the top two levels and a staff-level working group in the third. The cabinet-level principals of each branch of the Troika—chairman of the Council, director of the Office of Management and Budget, and secretary of the treasury—make up the highest level of the Troika, which is commonly referred to as T-1. T-2 generally includes the macroeconomic member of the Council, the associate director for economic policy from the Office of Management and Budget, and the assistant secretary of the treasury for economic policy. Staff members from the three agencies participate in the T-3 working group.

The Troika process usually begins about two months before the publication date for the budget document. After several preliminary meetings within each agency, consultation with outside experts in the economic and financial community, and considerable exploratory data analysis, the T-3 staff meets formally to develop a research agenda for the Troika concerning pertinent economic issues and draft a memorandum for T-2.

For the purposes of both forecasting and policymaking, each of the three branches of the Troika plays an important role. Economists at the Council work with their counterparts at the Department of the Treasury to provide background analysis for new policy initiatives, while the staff of the Office of Management and Budget works out the fiscal implications of the proposals. The Council of Economic Advisers' economic assumptions are used by both the Department of the Treasury and the Office of Management and Budget to produce budget revenue and outlay estimates, respectively. And at various times the principals from all three agencies participate in public debates concerning current policy issues.

The Troika's most important ongoing responsibility is to produce the administration's economic projections. These macroeconomic forecasts of any administration serve two purposes: (1) as a basis for the determination of the revenue and outlay estimates of the budget; and (2) as a policy statement by the administration.

> As a basis for the determination of the budget estimates, there is an implicit requirement for an honest and accurate assessment of current economic conditions in order to formulate reasonable assumptions on which to base the forecasts. At the same time however, because these forecasts are designed to reflect the presumably beneficial economic effects of the Administration's policy proposals, there is an inherent tension between developing "policy correct" economic assumptions which suggest a strong economy moving forward toward full employment with low inflation versus an honest assessment of current conditions and the likely impacts of these policies over the forecast horizon. The weights applied to these sometimes conflicting responsibilities can vary greatly as membership in the Troika changes. (Donihue and Kitchen 2000, p. 233)

Thus, the Troika forecast is partially subjective in nature, and this fact historically has led to the tension between accuracy and political purpose. Various statistical models and technical and theoretical relationships are used to construct the forecast, but they do not ultimately determine the final forecast.

This approach is a illustrative example of a combination of models and additional expert judgments, and is not very different from the methods used by the vast majority of macroeconomic forecasters, both public and private, although different incentives may exist. Judgment on the part of the forecasters typically enters in three ways: (1) in the specification of the model; (2) in determining forecast paths for the exogenous variables; and (3) in targeting, or "add-factoring" the near-term forecast values of the endogenous variables of the model to bring the model predictions in line with current data (see Donihue 1993). The process of generating a model-based forecast then becomes an iterative one of specifying alternative paths for the exogenous variables, fine-tuning the near-term forecasts of the endogenous variables, and checking the forecast throughout the horizon for reasonableness and consistency.

Donihue and Kitchen (2000) argue that there are a number of areas in which the "integrity" of the Troika forecast can be questioned. The administration's forecasts are policy forecasts, that is, the forecasts are made conditional on the adoption of the administration's policy proposals. "As a result, there is a political incentive toward positive bias in the economic projections which would reflect optimistic scenarios regarding productivity and employment objectives" (p. 239).

The case of the US Troika process shows that there is a desire among competing experts to come to a model-constrained consensus. The reason for this is that the experts know that a model-constrained consensus will be much more influential than if the policymakers are allowed to make their own choice from a set of conflicting experts' judgments. Where many modelers each have a different model based on a different theory, a kind of "colligation" is needed (den Butter and Morgan 1998). The relative relevance of the different models has to be judged and the different outcomes have to be framed—colligated—into one encompassing story, which is different from an encompassing model. A "rational" methodology for this kind of colligation will be explored in more detail in section 6.5.

6.3. The Delphi Method

To develop a methodology for inexact science that relies on expert judgments, Olaf Helmer designed at the Rand Corporation the "Delphi method," "undoubtedly the best-known method of eliciting and synthesizing expert opinion" (Cooke 1991, p. 12). The Delphi method is based on structural surveys and makes use of "the intuitive available information of the participants, who are mainly experts" (Cuhls 2005, p. 96). There is not one Delphi methodology; instead, there is a general agreement that it is an expert survey in two or more rounds in which in the second and later rounds the results of the previous round(s) are given as feedback. Thus, the experts offer answers from the second round onwards, taking account of the other experts' opinions that were offered in earlier rounds.

The Delphi method was devised in order "to obtain the most reliable opinion consensus of a group of experts by subjecting them to a series of questionnaires in depth interspersed with controlled feedback" (Dalkey and Helmer 1963, p. 458). The controlled feedback was the anonymous responses of the other experts and so designed to avoid "the disadvantages associated with more conventional uses of experts, such as round-table discussions or other milder forms of confrontation with opposing views" (p. 459). By means of this procedural feedback it was expected that the initially divergent views "will show a tendency to converge as the experimentation continues" (p. 459).

From its beginning, the Delphi method has three features:

- *Anonymous response*—opinions of members of the group are obtained by formal questionnaire.
- *Iteration and controlled feedback*—interaction is effected by a systematic exercise conducted in several iterations, with carefully controlled feedback between rounds.
- *Statistical group response*—the group opinion is defined as an appropriate aggregate of individual opinions on the final rounds. (Dalkey 1969, p. 408)

These features were designed to minimize the biasing effects of dominant individuals, of irrelevant communications ("noise") and of group pressure towards conformity. "In experiments at RAND and elsewhere, it has turned out that, after face-to-face discussion, more often than not the group response is less accurate than a simple median of individual estimates without discussion" (Dalkey 1969, p. 414).

In the middle 1960s and early 1970s the Delphi method found a broad variety of applications. Most applications are concerned with forecasting, but the method has also been applied to many types of policy analysis. We shall only be concerned here with the forecasting type of the Delphi method.

An example of an application of the forecasting type of the Delphi method is the forecast of oil prices by the California Energy Commission (Nelson and Stoner 1996). Since 1982, the commission has conducted surveys of oil price forecasts using the Delphi method as one of its forecasting tools. The Delphi method was chosen because it incorporates a number of features that make it particularly attractive. The Delphi method is especially useful for long-range forecasting (20–30 years). The method is flexible, imposing a common response format on panelists, and allows consistent, systematic statistical treatment. The method makes it possible to consult with and incorporate the views of a relatively large number of (e.g., geographically) dispersed experts. The anonymity guaranteed to panelists encourages candor.

In forecasting, the team conducting the Delphi method seeks experts who are most knowledgeable on the issues in question, and seeks to achieve a high degree of consensus regarding predicted developments.[4] Its basic approach can be described as follows (see Cooke 1991, pp. 13–14): A monitoring team defines a set of issues and selects a set of respondents who are experts on the issues in question. A respondent generally does not know who the other respondents are, and the responses are anonymous. A preliminary questionnaire is sent to the respondents for comments, which are then used to establish a definitive questionnaire. This questionnaire is the sent to the respondents and their answers are analyzed by the monitoring team. The set of responses is then sent back to the respondents, together with the median answer and the interquartile range, the range containing all but the lower 25% and the upper 25% of the responses. The respondents are asked if they wish to revise the initial predictions. Those whose answers remain outside the interquartile range for a given item are asked to give arguments for their prediction on this item. The revised predictions are then processed in the same way as the first responses, and arguments for outliers are summarized. This information is then sent back to the respondents, and the whole process is iterated. A Delphi exercise typically involves three or four rounds. The responses on the final round generally show a smaller spread than the responses on the first round, and this is taken to indicate that the experts have reached a degree of consensus. The median values on the final round are taken as the best predictions.

Cooke (1991, p. 15) notes that one of the most important later variations of the Delphi method was the involvement of letting the experts indicate their own expertise for each question. Only the judgments of the experts claiming the most expertise for a given item are used to determine the distribution of judgments for that item. It was claimed to improve accuracy. This claim, however, was challenged by a study by Klaus Brockhoff ([1975] 2002) showing that self-ratings of participants did not coincide with "objective expertise" as measured by relative deviation for the true value on fact-finding and forecasting tasks: "Self-ratings of expertise show a positive relationship to the performance of the persons questioned in only two of four Delphi groups. . . . It is important to employ and develop better methods for the determination of expertise" (Brockhoff [1975] 2002, p. 311).

6.4. Rational Consensus

According to Roger M. Cooke (1991), the kind of consensus reached by the Delphi method is not "rational": Delphi consensus does not imply convergence or accuracy, or more generally is not "scientific." Therefore, he proposed five "principles" to formulate "guidelines for using expert opinion in science" (p. 81):

- Reproducibility. "It must be possible for scientific peers to review and if necessary reproduce all calculations. This entails that the Calculational models must be fully specified and the ingredient data must be made available" (p. 81).
- Accountability. "The source of expert subjective probabilities must be identified" (p. 81).
- Empirical control. "Expert probability assessments must in principle be susceptible to empirical control" (p. 82).
- Neutrality. "The method for combining/evaluating expert opinion should encourage experts to state their true opinions" (p. 83).
- Fairness. "All experts are treated equally, prior to processing the results of observations" (p. 83).

He concluded the discussion of these "rational" principles for expert consensus with the remark, "There is no method at present which satisfies all of the above principles. There is every reason at present to develop these methods" (p. 84). And so he did. Over the last 20 years, at Delft University of Technology, Cooke has developed procedures to support the formal application of expert judgment to acquire "rational consensus." These procedures are set out in a *Procedures Guide for Structured Expert Judgment* (Cooke and Goossens 1999), which was developed for a uncertainty study of accident consequence codes for nuclear power plants using structured expert judgment, commissioned by the European Commission and the United States Nuclear Regulatory Commission.

In this guide the "principles for rational consensus" (p. 15) were slightly revised, into the following:

- Scrutability/accountability: all data, including experts' names and assessments, and all processing tools are open to peer review and results must be reproducible by competent reviewers.
- Fairness: experts are not prejudged.
- Neutrality: methods of elicitation and processing must not bias results.
- Empirical control: quantitative assessments are subjected to empirical quality controls.

This revision implies a shift of emphasis in the principles of fairness and neutrality: from aiming at unbiased judgments from the individual experts to aiming at a method of combining the individual expert judgments such that the "consensus" is unbiased.

Unlike the Delphi method, rational consensus is not based on an agreement of the experts. "If rational consensus requires expert agreement, then rational consensus is simply not possible in the face of uncertainty"; even if "quarantined," the experts would disagree (p. 15). The proposed rational consensus is a weighted average of the individual expert judgments. Thus, the main problem is the determination of these weights.

In a survey article discussing 15 years of research in expert judgment at Delft, additional considerations are given for preferring a rational consensus, that is, a "mathematical aggregation" above an "agreement among experts":

> A group of experts tends to perform better than the average solitary expert, but the best individual in the group often outperforms the group as whole. . . . This motivates the elicitation of the assessments of individual experts without any interaction, followed by mathematical aggregation in order to obtain a single assessment per variable, thereby weighting the individual experts' assessments based on their quality. (Goossens et al. 2008, pp. 234–235)

Over the years, the term "rational consensus" has come to receive a more specific meaning, namely "mathematical aggregation," that is, a weighted aggregation, where the judgment of each expert is weighted on the basis of his or her knowledge and ability to judge relevant uncertainties. Again, an essential element of the method of rational consensus is the weighting of the experts. This element will be discussed in what follows.[5]

The method, also called the "Cooke method," has found successful applications for cases where one cannot build up historical data to quantify risk models and has proven to be "most effective when data are sparse, unreliable or unobtainable" (Aspinall 2010, p. 294). For these cases,

> Expert judgment is used to obtain results from experiments and/or measurements, which are physically possible, but not performable in practice. Such experiments are "out of scale" financially, morally, or physically in terms of time, energy, distance, etc. They may be compared to thought experiments in physics. Since these experiments cannot in fact be performed, experts are uncertain about the outcomes, and this uncertainty is quantified in a formal expert judgment exercise, so as to obtain the information that we require. (Cooke and Goossens 1999, p. 24)

In a video interview made at Resources for the Future, Cooke expressed this as follows:[6]

> We never built up historical data to quantify our risk models by the nature of the case. So in building a risk model and quantifying this model we must have recourse to expert judgment, and this has been a theme throughout my work, and a lot of risk analysis is directed to this. We really want to look at expert judgment as a new form of scientific data which we can use in a methodological proper way to quantify our models and make this whole process transparent. (Cooke 2009, transcription by the author)

Instead of performing a physical experiment, experts are asked to do a "hypothetical experiment":

> First you have to be very clear what you want to ask, and this is a very difficult part of an expert judgment exercise to formulate a protocol of the questions you exactly . . . exactly what you want to know. I like to think of it as follows: that expert judgment is just a different way of doing experiments. And you should have questions that you could in principle ask nature, but for various reasons practically you cannot do so. So you ask these questions to experts who are familiar with the whole field and who can tell you what they think will probably happen, and what their uncertainty is on what is going to happen if you could do such an experiment. (Cooke 2009, transcription by the author)

If a parameter is uncertain, and if the uncertainty cannot be quantified with historical and/or measurement data, then the analyst must ask the expert how the values *would* be determined if suitable measurements could be performed. Although these experiments are hypothetical, that is, they cannot be performed in practice, they must be physically possible. The values are known to depend on a large number of physical parameters that cannot all be measured or controlled in any given experiment. Moreover, the functional form of the dependence is not known. Hence, if a controlled experiment is repeated many times, different values will be found reflecting different values of uncontrolled and unknown physical parameters. If a measurement setup is described to an expert, he or she can express uncertainty via a subjective distribution over possible outcomes of the measurement. In such cases the experts are questioned directly about uncertainty with respect to model parameters. In other words, experts are "elicited" to give a Type B uncertainty evaluation (see chapter 2).

The Cooke method is thus an approach where experts are being purposely used to run imaginary experiments; that is, the Cooke method is a combination of the approaches of Helmer (expert elicitations) and Hempel (imaginary experiments).

A formal expert judgment exercise is called an "elicitation." And because this exercise is to reveal and quantify the expert's uncertainties, Cooke considers the preparation for elicitation as "really nothing more than carefully designing these hypothetical experiments, so as to obtain the information that we require" (Cooke and Goossens 1999, p. 24).

> And then we look at experts really as statistical hypotheses. When an expert says he has a certain uncertainty distribution over the range of outcomes of some possible experiments: that is a statistical hypothesis. And that is how we look at it. (Cooke 2009, transcription by the author)

In describing this hypothetical experiment to the expert, the physical factors that may influence the outcome of the experiment are first identified by an "analyst." Each relevant physical factor will fall into one of the two classes: (1) the case structure assumptions; and (2) the uncertainty set. Some relevant factors will have their values stipulated by the assumptions of the study, as reflected in the case structure. Other factors may influence the outcome of the hypothetical experiment, but their values are not stipulated by the case structure. These factors belong to the uncertainty set. The experts are made aware that these factors are uncertain, and should fold this uncertainty into their distributions on the outcome of the hypothetical experiment. The general format for elicitation is given in figure 6.1.

An important phase of the procedure is the "identification" of the expert. An "expert for a given subject" is defined as a "person whose present or past field contains the subject in question, and who is regarded by others as being one

Conditional on

< values of factors in the case structure assumptions >

Please give the 5%, 50% and 95% quantiles of your uncertainty in

< Hypothetical experiment >

taking into account that values of

< uncertainty set >

are unknown

Figure 6.1 Format for Eliciting Continuous Variables. Source: Figure 5, Cooke and Goossens 1999, p. 27.

of the more knowledgeable about the subject" (pp. 29–30). These experts are designated as "domain" or "substantive" experts. The following general selection criteria are recommended (p. 30):

- Reputation in the field of interest
- Experimental experience in the field of interest
- Number and quality of publications in the field of interest
- Diversity in background
- Awards received
- Balance of views
- Interest in and availability for the project

These selection criteria are probably chosen to meet the rational consensus principle of scrutability/accountability.

Once the experts are selected, they are requested to provide their assessments on the query variables. The next crucial phase is the determination of the weight of these assessments. It is for this determination that the last "principle for rational consensus," "empirical control," comes into play. Empirical control is built into the elicitation procedure by asking experts to assess "calibration" or "seed" variables. "Seed variables" are variables whose values are or will be known to the analyst within the frame of the exercise but not to the expert. "Seed variables are important for assessing the performance of the combined experts' assessments. Seed variables also form an important part of the feedback to experts, helping them to gauge their subjective sense of uncertainty against quantitative measure of performance" (Cooke and Goossens 1999, p. 28).

Although it is explicitly noted that expert assessments should not be treated "as if they were physical measurements in the normal sense, which they are not" (p. 10), they are assessed as if they are measuring instruments, namely by calibration. Calibration and gauging are typical techniques for increasing the reliability of a measuring instrument, but here they are applied to expert judgments: "expert judgment is recognized as just another type of scientific data, and methods are developed for treating it as such" (p. 10).[7]

Empirical control of the expert's performance is used to determine the weights of the expert's judgments in the aggregation of them. This performance is measured by the expert's assessment of "seed variables." As Cooke and Goossens (1999, p. 12) explain, seed variables may sometimes have the same physical dimensions as the variables of interest. This arises when the variables of interest are not practically measurable for reasons of scale, for example, great distances, long times, high temperatures; whereas measurements can be performed at other scales. In this case, unpublished measurements or experiments can be used as seed variables. When such seed variables are not available, variables can be chosen that "draw on the relevant expertise" yet do not have the same dimensions as the variables of interest. As a loose criterion, a seed variable should be a variable about which the expert may be expected to make an educated guess, even if it does not fall directly within the field of the study at hand.

The identification of appropriate seed variables is a difficult task of the "uncertainty analyst": "It is impossible to give an effective procedure for generating meaningful seed variables. If the analyst undertakes to generate his own seed variables, he must exercise a certain amount of creativity, perhaps supported [by] the experts themselves" (Cooke and Goossens 1999, p. 28).

Seed variables falling squarely within the experts' field of expertise are called "domain variables." In addition to domain variables, it is permissible to use variables from fields that are adjacent to the experts' proper field. These are called "adjacent variables." Adjacent variables are those about which the expert should be able to give an educated guess. Seed variables may also be distinguished according to whether they concern predictions or retrodictions. For predictions the true value does not exist at the time the question is answered, whereas for retrodictions, the true value exists at the time the question is answered, but is not known to the expert. Cooke and Goossens (1999) provide the following evaluation of these four different kinds of seed variables:

> In general, domain predictions are the most meaningful, in terms of proximity to the items of interest, and are also the hardest to generate. Adjacent retrodictions are easier to generate, but are less closely related to the items of interest. (Cooke and Goossens 1999, p. 29)

Combining these notions, they summarize "a crude evaluation" of the four types of seed variables in a table (table 6.1).

Table 6.1 **Classification of Seed Variables, Crude Evaluation**

	Predictions	*Retrodictions*
Domain	+ + +	+ +
Adjacent	+ +	+

Source: Cooke and Goossens 1999, p. 29.

6.5. Model-Based Forecasting

Both the Delphi method and the Cooke method were developed by mathematicians and engineers specifically for applications in relation to engineering problems. This impedes the application of the Cooke method to the typical problems in social science. The identification of appropriate seed variables concerning retrodictions is more difficult in social science than in natural science for reasons like those relevant to the application of Type B uncertainty evaluations in social science, as discussed in chapter 2. The use of seed variables for calibration purposes presumes scientific consensus on their true values. In social science it will be very hard to find variables for which there is scientific consensus and at the same time are unknown to the expert.

This does not hold for seed variables concerning predictions. By their nature they are unknown to the expert when he or she is solicited about them. Moreover, in Cooke and Goossens's "crude evaluation" table, domain predictions appear as the strongest seed variables; see table 6.1. Therefore, for discussion of the involvement of expert judgments in social science we now focus on "judgmental forecasting."

In forecasting, it is acknowledged more explicitly than in other fields that besides a model one also needs judgment, despite the more recent downgrading of expert judgment in various other fields (see chapter 5). In a special issue of the *International Journal of Forecasting* on 25 years of forecasting, the importance of the role of expert judgment was clearly acknowledged:

> While judgement has always played an important role in forecasting, academic attitudes to the role and place of judgement have undergone a significant transformation in the last 25 years. It is used to be commonplace for researchers to warn against judgement . . . , but there is now an acceptance of its role and a desire to learn how to blend judgement with statistical methods to estimate the most accurate forecasts. The forecasting practitioner has never shared the scepticism of the researcher towards judgement. (Lawrence et al. 2006, p. 493)

In the Cooke method, expert consensus presupposes consensus on the values of the seed variables. This consensus is crucial to the method because these values are used to calibrate the expert judgment, that is, to give the expert judgment a "weight." In a science like economics, values of variables do not and will never have this decisive role, because their supposed values will have too much uncertainty. In economics, for example, it is not customary to talk in terms of "true values," but instead of "estimates" or "approximations."

In economics, one will find much more consensus on the validity of specific empirical models than on values of variables, in particular when these models are developed at recognized economic institutions, like central banks. Therefore, to enable the application of the Cooke method to social science, I suggest a shift from consensus on specific facts to consensus on specific empirical models. Cooke's rational-consensus method does not optimally use this shared knowledge captured by models. According to the rational-consensus procedures, models are considered to be more personal. The individual experts can bring to bear "their written rationales," these including "consult of sources, do some modeling, do some calculations, run some codes," but these "rationales" are personal and not necessarily shared intersubjective knowledge (Cooke and Goossens 1999, p. 33). That may be true for engineering, but in economics there is much more consensus on the validity of the models used at, for example, central banks. The weighing of experts should therefore not be based on calibration using seed variables, but instead should be based on calibration with seed models.

In a recent article in the *International Journal of Forecasting*, Philip Hans Franses (2008, p. 31) called for "more interaction between researchers in model-based forecasting and those who are engaged in judgemental forecasting research." His call is built on some positive evidence on the successful interaction between models and experts. According to Franses, there seems to be sufficient evidence that combining models with expert judgments leads to better judgments. This seems even more evident with respect to forecasts. "Indeed, a model's forecast, when subsequently adjusted by an expert, is often better in terms of forecasting than either a model alone or an expert alone" (Franses 2008, pp. 32–33). The reason is that "a model will miss out on something, and that the expert will know what it is" (p. 32).

In the context of the use of economic models for forecasting purposes, the reason for having additional involvement of expert judgments is similar to the reason for using the Cooke method mentioned above, namely, to correct for obvious known shortcomings in the economic model or to mimic the effects of economic events occurring outside the model:

> Shortcomings can occur when actual time series do not fit well with the estimated behavioural equation, for example because of revisions of the national accounts. Outside economic effects can involve specific knowledge for the near future about contracts or plans or the creation of temporally higher or lower effects of economic behaviour of households or firms because of sudden shocks in confidence or announced changes of tax rates. (Franses, Kranendonk, and Lanser 2007, p. 7)

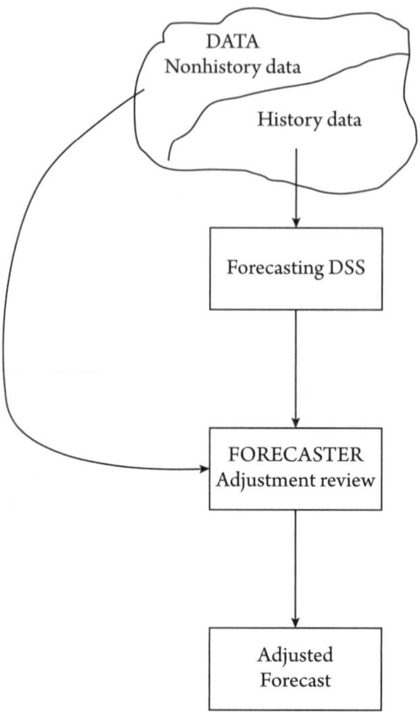

Figure 6.2 Forecasting Steps. Source: Lawrence et al. 2006, p. 494.

In model-based forecasting, expert judgments can be used in two different ways, namely, by interfering with the specification of the model structure or by adjusting "add factors." An add factor is an additional factor in the equations of the model concerning the behavior of households and firms. The add factor can be used by the forecaster to adjust the outcome of the equation.

Published forecasts are rarely based on the model outcome only; additional adjustments are often made to the model-based predictions in arriving at a final forecast. Thus published forecasts reflect in varying degree the properties of the model and the skills of the model's proprietors. This will be illustrated by two figures.

The first figure, here figure 6.2, is presented in a discussion by Michael Lawrence, Paul Goodwin, Marcus O'Connor, and Dilek Önkal (2006) on 25 years of "judgmental forecasting" and pictures the essential steps in forecasting. The figure is explained in terms of a case on the sales of a product:

> We propose viewing the total set of data useful for forecasting as made up of two classes; the history data and the domain or contextual data. The history data are the history of the sales of the product. The domain data are in effect all the other data which may be called on to help understand the past and to project the future. This includes past and future promotional plans, competitor data, manufacturing data and macroeconomic forecast

> data. The data usually input to a forecasting decision support system are the history data and occasionally promotion data. The adjustment review process is informed by both the history data and all the non-history data. (Lawrence et al. 2006, p. 494)

Judgment is needed to account for the "non-history data."

The second figure, here figure 6.3, was developed to clarify macroeconomic forecasting of the CPB Netherlands Bureau for Economic Policy Analysis. In January 2011, the bureau organized two meetings to inform journalists and policymakers about the models being used at the bureau and how forecasts are made, under the title "Models and Forecasting: A Look in the Engine Room of the CPB." One key message was that the CPB does not blindly trust computer calculations. The results are always assessed by experts on subfields (Verbruggen 2011).

Although it is already a long-standing practice in forecasting to have expert judgments involved, there is, however, still a lack of studies concerning the "weighting" of the individual experts, that is, the "empirical control" of the expert judgments to enhance "rationality." Lawrence and coauthors (2006, pp. 506–510), however, discuss several strategies for improving judgmental forecasts, of which two can be considered as possible for empirical control of expert judgments in social science.

The first improvement strategy they discuss is the provision of outcome feedback. This strategy, however, suffers the same problem as seed variables concerning retrodiction, namely, "most recent outcome contains noise and hence it is difficult for the forecaster to distinguish the error arising from a systematic deficiency in their judgement from the error caused by random factors" (p. 507).

A second strategy that may be used for weighting expert judgments is the use of statistical forecasting methods to forecast the error in judgmental forecasts. The predicted error can then be used to correct the judgmental forecast.

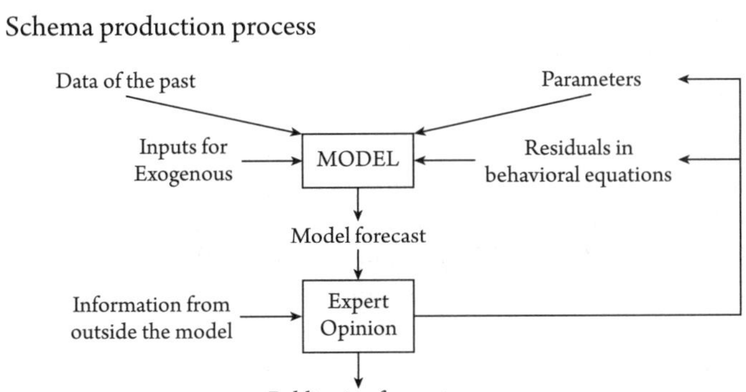

Figure 6.3 Scheme of the Production of a Published Forecast. Source: Translated from Verbruggen 2011, slide 6.

This correction method is appropriate when the biases associated with judgmental forecasts are systematic and sustained, for example, a tendency of forecasters to overweight recently released information (p. 509). Because this method presumes a longer record of judgmental forecasts, it is not as applicable to an individual expert as to aggregated judgmental forecasts. The disadvantage of this method is that it leads us back to the earlier problem of how to aggregate expert judgments rationally.

This latter problem, however, can be evaded by taking the "expert" not an individual scientist but a specific scientific institution, where a team of experts is employed. In the Delphi method and the Cooke method, the individual experts have a dominant role, but this came to be so because of the nature of the cases where these methods were applied. For these cases, there is generally no strong consensus about which model would be the most appropriate. Moreover, although expert judgments are needed to make a model complete, this requirement does not say who or what the carrier of this expert knowledge should be. My suggestion is that to deal with issues in social science, the carrier of expert knowledge is a scientific institution. To be scientific, an institution has to meet the same criteria as an individual expert, like the ones listed by Cooke and Goossens (1999); see section 6.4.

If we look at judgmental forecasts made by institutions and not by individual experts, we can use the previously mentioned second strategy of weighing expert judgments as an appropriate empirical control of scientific institutions. A good example of such a calibration is a study by Franses, Henk Kranendonk, and Debby Lanser (2011), in which they evaluated the macroeconomic forecasts of the CPB Netherlands Bureau for Economic Policy Analysis for the period 1997–2008.

As previoulsy mentioned, the published forecasts of the CPB are never simply the model forecasts. Before being published, all forecasts are scrutinized by various experts who assess the accuracy and adequacy of the model forecast and suggest judgmental adjustments. It is these adjusted forecasts that are made publicly available. Fortunately, and "in complete contrast to other forecasting areas" (Franses, Kranendonk, and Lanser 2011, p. 483), the CPB has kept track, at least since 1997, of the nature and size of these judgmental adjustments. Franses, Kranendonk, and Lanser could therefore use this database to compare the accuracy of the model forecasts with that of the judgmentally adjusted forecasts, to know how effective and useful these adjustments are.

Their main findings are that the CPB model forecasts are inaccurate, or "biased," for a range of variables. But at the same time the associated expert forecasts, that is, the outcomes of the model filtered by experts, are more often accurate; and expert forecasts are far more accurate than the model forecasts, in particular when the forecast horizon is short. In summary, "the final CPB forecasts de-bias the model forecasts and increase the accuracy relative to the initial model forecasts" (p. 494).

6.6. Conclusion: Model-Based Consensus

The idea of a model-based consensus is not new. It was originally conceived by the inventor of macroeconometric models, the Dutch economist Jan Tinbergen. He built the first two macroeconometric models, the first of the Dutch economy (Tinbergen 1936a) and the second of the US economy (Tinbergen 1939b). His method of econometric modeling was new and controversial at that time (see Morgan 1990 for an extensive history of these models and the debates to which they led). Therefore, while working on his second model for the League of Nations, Tinbergen wrote a memorandum to explain and justify this new method. The "method" he had developed aimed at understanding the causation of business-cycle phenomena, and "essentially starts with a priori considerations about what explanatory variables are to be included. This choice must be based on economic theory or common sense" (Tinbergen 1939b, p. 10). He was quite aware of the fact that economists do not agree upon which are the most important causes of the business-cycle phenomenon. The question then is how to deal with this disagreement. His solution was to integrate these different views into one model and to estimate the significance of each view:

> It is rather rare that of two opinions only one is correct, the other wrong. In most cases both form part of the truth. . . . The two opinions, as a rule, do not exclude each other. Then the question arises in what "degree each is correct"; or, how these two opinions have to be "combined" to have the best picture of reality. [We can] combine these different views, viz. by assuming that the movements . . . can be explained by some mathematical function of all the variables mentioned. We then have not a combination in the physical sense—an addition of two quantities or of two amounts—but a combination of influences. In many cases the mathematical function just mentioned may be approximated by a linear expression. (Tinbergen 1936b, pp. 1–3)

In a later recollection on this period, Tinbergen (1979) explained that he had learned this method from his mentor and Ph.D. thesis supervisor, the physicist Paul Ehrenfest. Ehrenfest had taught him,

> to formulate differences of opinion in a "nobler" way than merely as conflicts. His favourite formulation was cast in the general form: if $a > b$, scholar A is right, but if $a < b$, then scholar B is right. The statement applied to a well-defined problem, and both a and b would generally be sets of values of elements relevant to the problem treated, with possibly a number of components of qualitative nature. (Tinbergen 1979, p. 331)

This view on models and on how to use them in dealing with differences of opinions would never leave him. In a 1982 article "The Need of a Synthesis," Tinbergen repeated his lifelong credo (see also one of this chapter's epigraphs):

> In quantitative terms one can also say that in certain regression equations some of the coefficients indicate what the weights are of the explanatory variables, as these are put forward in the competing theories, in the explanation of the independent variable. In the search for a synthesis what matters is that, as has been stated by Klein: "It is less important that the effort be labeled Keynesian, monetarist, neoclassical or anything else, than that we get good approximation to explanation of this complete system. . . ." Indeed, the criterion by which we test the various competing views, has to be in the best possible explanation of the developments observed. . . . The point with which I want to end this argument is that the synthesis is only completed when such partial studies—the usefulness of which I accept fully—are made part of a complete model. The reason for that I gave earlier already: consistency with the other "blocs" of a complete model. That is why we cannot do without our largest model factory, the Central Planning Bureau, in establishing the synthesis intended. (Tinbergen [1982] 2003: 303, 305–306)

Tinbergen saw models as means to reach consensus. "They make it possible to localize differences of opinion: to indicate the equation about which one disagrees, the term of that equation, or the term that is lacking, or the variable that is lacking" (Tinbergen 1987, p. 106; translated by the author).

Tinbergen was the first director of the CPB Netherlands Bureau for Economic Policy Analysis, founded in 1945, and became as such one of the most important designers of Dutch consensus-based policy, also called the "Polder model." Today, economic policy analysis at the CPB is still very much in the tradition of its first director:

> Many policy measures in the macroeconomic sphere can only be understood and discussed properly with the help of a model which sets out the key relationships between macroeconomic variables. Such a model is an important instrument in considering relevant relationships. (Don and Verbruggen 2006, p. 146)

In a panel discussion to explore policymakers' perspectives on their experiences of the modeling-policy interaction, Don explained the unique role models have in Dutch policymaking, when compared to other countries:

> Perhaps the most important one is to use models as an information processing device: to monitor the economy, to monitor the budget outlook

> in particular, and to provide information about different scenarios that the near future might bring. In Dutch policy-making there is still another use of economic models, which is to use it as a tool in consensus building. . . . Using the model may help to locate exactly where the political differences are and whether these are differences of preferences in what people would like the economy to produce or whether these are difference in analysis of what the economic trade-offs really are. The model helps very much in assessing all these difference and in getting as much common ground as you can get. (Don quoted in Morgan 2000, pp. 264–265)

In this Tinbergenian kind of model-based consensus, however, the role of the expert is actually as modest as the role of the "economist" in Koopmans's "logic of econometrics," discussed in chapter 4.[8] The economist identifies the set of potential influences, whose magnitudes then will be measured by the statistician to see whether they are significant in the causal explanation of the business cycle, for example.

According to Tinbergen's and Koopmans's account, the "additional information" of the expert can be measured statistically, that is, estimated, and the assessment of this information therefore is a Type A evaluation. The problem remains that we also need experts to add nonmeasurable information, that is, information for which we do not have statistics. For this kind of information, the role of the expert judgment is to provide a "measurement" when no statistical measurement is possible. The evaluation of this latter kind of expert judgment cannot be conducted by measurement as Tinbergen and Koopmans proposed. It should be a Type B evaluation.

As we saw in the preceding section, the empirical control of expert judgment by using predictions as seed variables is quite problematic because of the uncertainty of the data themselves. Experts in social science are not validated by their predictive performance: even a forecaster that had it right at a particular prediction does not come to be considered as a recognized expert; he or she still could have had it right by accident.

In chapter 2 we arrived at a model-based Type B evaluation. To apply this kind of evaluation to "control empirically" expert judgments, expert knowledge should be considered a black-box model and therefore should be validated accordingly, that is, by behavior pattern tests.[9] Because the "imaginary model" of an expert is "intuitive," it makes no sense to validate its structure. Using Reichenbach's distinction between "context of discovery" and "context of justification," the "context" of how the judgments are made by the expert, that is, the application of his or her imaginary model, is not relevant for the justification of the expert's judgment.

The behavior pattern tests, however, should not be limited to point predictions. They should include a broader range of predictions about behavior patterns, like frequencies, trends, phases, lags, and amplitudes. An expert on volcanos, a volcanologist, is not expected to predict the timing of a volcanic eruption, but instead to be able to make an adequate analysis of the observed changes in certain conditions and based on this analysis to make an adequate prediction about a probable eruption with a probable magnitude within a probable period.[10] Experts who in the past regularly made good predictions of trends and turning points should gain a higher "weight" in the aggregated judgment.

Notes

1. See Martini 2014 for a related account on experts and consensus. When I presented an earlier version of this chapter at the Ninth Conference of the International Network for Economic Methodology in Helsinki (September 2011), I discovered that Carlo Martini shared the same interest in experts and consensus, and was studying the same literature. This led to a fruitful cooperation that resulted in a volume *Experts and Consensus in Social Science* (2014), edited by Martini and me.
2. Jick's discussion of divergent results looks similar to Whewell's method of residues (see section 3.3), but he does not refer to Whewell.
3. The discussion of the Troika process is based on Donihue and Kitchen 2000.
4. See Cooke's (1991) *Experts in Uncertainty* for a more detailed discussion of the Delphi method. The brief exposition of this method here is based on Cooke's (1991) account.
5. Cooke was not, however, the first one who designed procedures for reaching "rational consensus." In the late 1970s, Keith Lehrer and Carl Wagner (1981) developed a mathematical model of consensus that was also based on a weighted aggregation of individual probability assignments. The main difference with Cooke's method is that the weights are not determined by measured performance but by the "opinions" experts have about each other: "Our method for finding rational consensus rests on the fundamental assumption that members of a group have opinions about the dependability, reliability and rationality of other members of the group" (p. 19). I thank Rainer Hegselmann for bringing this earlier work on rational consensus to my attention.
6. Since September 2005 Cooke has been appointed to the Chauncey Starr Chair in Risk Analysis of Resources for the Future.
7. Although experts are not considered to be measuring instruments, their judgments are considered as measurements; see also chapter 5 for a discussion of the "symmetrical roles" of rational judgment and measurement.
8. Which is not surprising, because Koopmans's econometric methodology was highly influenced by Tinbergen. Tinbergen, however, made one exception to the rule that an expert judgment should not be considered as scientific data (unlike the Cooke method), namely, in the case when the expert is no less than Keynes: "Sometimes, indeed, intuition constitutes a basis for new scientific results. It should be the intuition of a genius, however. For simpler souls, intuition may be less reliable!" (Tinbergen 1979, pp. 342–343).
9. While I see expert knowledge as models, Erika Mansnerus (2014), in her account of modeling in epidemiology, considers models as "senior experts."
10. This analogy with volcanology is made in a CPB report (de Jong, Roscam Abbing, and Verbruggen 2010, p. 61) to underline the report's main conclusion: "due to the character of macro-economic short-term forecast, it is most unlikely that CPB and other forecasting institutes will be able to forecast the next financial crisis adequately" (p. 3). The report was written to answer the question why CPB "did not see the credit crisis coming, nor did it predict that the Dutch economy would shrink in 2009" (p. 3). The presentation "Models and Forecasting" by Verbruggen, discussed in section 6.5 (see also figure 6.3), was based on this report. The report thus discusses explicitly the necessary role of experts.

7

Conclusions

> Reports that say that something hasn't happened are always interesting to me, because as we know, there are known knowns; there are things we know we know. We also know there are known unknowns; that is to say we know there are some things we do not know. But there are also unknown unknowns—the ones we don't know we don't know.
>
> —Donald H. Rumsfeld (2002)[1]

Studying measurement is studying science. Measurement is the assignment of numbers to a property of a phenomenon, called a measurand, according to a rule. This book has analysed what these rules are to have these numbers provide reliable knowledge about the measurand.

The measurement accounts that have been presented and discussed in this book are only a small selection of the many practices one will find in social field science and which all deal with the same problem: how to get reliable measurement results when one cannot apply the experimental method of the laboratory. Because these accounts reflect practice, many of them have to be inferred from almanacs, dictionaries, guides, handbooks, instructions, reports, teaching materials, tutorials, and yearbooks.

The assessment of the reliability in a field science is different from the assessment of the reliability in a laboratory science. A field science studies phenomena that cannot be studied in a laboratory for practical, technical, or ethical reasons, which means that these phenomena cannot be isolated from their environment and cannot be investigated by systematic manipulation or intervention enabling the measurement results to be reproducible. But it is not science in the "wild," that is, measurement of a field phenomenon that is planned and designed, which often entails the instalment of institutions, networks, or organizations that enable field study. Nevertheless, these institutions are different from laboratories; that is, institutions of field science are not exclusively scientific domains.

Because measurement of field phenomena is not possible by instantiating intervention and control in a scientific environment, and so is based on passive field observations, it requires a different methodology, that is, a methodology whose rigor is not based on the standards of the laboratory but on standards appropriate for the specific field. While the standards of the laboratory are "universal"—it does not matter where the laboratory is equipped—the standards of the field are much more context dependent and sensitive to local conditions. But less universal does not have to mean less rigorous.

This methodological tension between laboratory science and field science is traditionally expressed in terms of a contrast between the standards of natural science and those of social science. For example:

> The contrast . . . between the experimental method of the natural sciences and the statistical of the social sciences may be expressed in greater detail. The natural scientist sets up and tries his experiment; repeats it as often as he pleases under the same or varying conditions; isolates the factor in which he is interested; and arrives at a demonstrable conclusion concerning the operation and effects of that factor. The social scientist, on the other hand, must accept and analyse the mixed situation as it comes to him; gather pertinent statistics, not such as he would like, but such as are available; study figures which embody the combined effects of many factors; and express his conclusions in terms of probabilities. (Persons 1924, pp. 2–3)

While laboratories are exclusively scientific domains, the field is public space populated not only by scientists but also by people with other interests. Social statistics are not the result of designed experiments; they are often dependent on institutional rather than scientific definitions of processes. Moreover, even when planned, social statistics are generally not gathered by scientists. Because field observations are usually collected by nonscientific institutions and organizations with other than scientific interests, one has to account for an additional problem, namely, deliberate deceit.

Field observations are also more personal than laboratory observations, particularly in the case of the observation of an unique event. To make such an observation objective, that is, credible and plausible, one cannot rely on the trust-inducing procedures of the laboratory, like reproducibility, and so the credibility of the observer—in addition to the reliability of the laboratory, instrument, or model—has to be taken into account.

There is another relevant difference between a laboratory science and a field science: A field science is much more inexact than a laboratory science. In an exact science the models are complete, that is, the model comprises all large and systematic causal factors; all remaining influences can be treated as a noise component that can be modeled as a draw from some probability distribution. Hence, exact science presuppose complete knowledge, that is, that we know all relevant causal factors. But the problem for a field science is that there are unmeasurable knowns, known unknowns, but also unknown unknowns, which can be large and so cannot be stowed away in the noise component.

To study field phenomena, the laboratory has been replaced by a statistical model. Instead of experimenting on material phenomena, one can experiment on the virtual phenomena existing in the world of the model.

If the theories about a specific phenomenon can be expressed in an exact and complete model, then the experimentation in the "world of the model" (Morgan 2012) can be run on a computer—usually called simulation. But the problem of our knowledge about field phenomena is that our theories are often incomplete and inexact. The incompleteness and inexactness of our theories have to be complemented by more intuitive knowledge. For these cases, the experimentation cannot be run on a computer, but instead has to be conducted in the laboratory of the mind—usually called a thought experiment. The subsequent problem is to decide whose intuitions are most appropriate, or in other words, who is best equipped to run the thought experiment. In other words, one has to decide who the experts are that are needed to complement our incomplete theories.

In the most prominent current measurement theory, the representational theory of measurement, the models have to be homomorphic mappings of the measurand, that is white-box models. It has been shown in this book that gray-box and even black-box models can also be used for measurement purposes; it is only that the validation has to be different. Different, however, does not necessarily mean less rigorous. The advantage of a gray-box model is that our knowledge of the measurand need not to be exact and complete.

A measurement expresses—in a quantitative way—what we know about a certain phenomenon and what we do not know about it. The measurement value expresses what we know about the measurand; the reliability of the measurement value expresses what we do not know, how uncertain we are about the validity of the measurement value. Measurement, therefore, consists of two interconnected aspects, the known and the unknown about the measurand, both expressed in numbers. Both aspects are determined by a methodology called a "calculus of observations." This calculus of observations entails not only the employed procedures, tools, techniques and methods, but also the assessments of these.

A calculus of observations is a "clinical judgment" in the sense that it is partly objective and partly subjective. The objective part consists of mechanical procedures for calibration, acquiring precision, and Type A evaluations, and is traditionally taken as capturing the whole measurement. The subjective part consists of the choice of the model, the choice of the standards, and Type B evaluations. These latter judgments are based on expert knowledge, that is, knowledge of an expert based on professional skill, training, and experience with the measurand.

Field observations are designed and planned observations, that is, observations provided by institutions. Usually these are called "statistics." But "statistics" has two meanings. Maurice Kendall and Alan Stuart (1963) define statistics as "the branch of scientific method which deals with the data obtained by counting or measuring the properties of populations of natural phenomena. In this definition 'natural phenomena' includes all the happenings of the external world, whether human or not" (p. 2). The same word "statistics" is also applied to the numerical material with which the method operates, so the data obtained by counting and measuring. More generally, statistics in the sense of data is defined as quantitatively registered history. This registration is generally carried out by institutions.

Is statistics, that is, method and data combined, sufficient for acquiring knowledge about the deeper structure of a phenomenon? This question was explicitly studied by Haalvelmo: "Can we measure economic structure relations (e.g. individual indifference surfaces or other 'behavioristic' relations) by means of data which satisfy simultaneously a whole network of such relations, i.e., data obtained by a 'passive watching of the game' and not by planned experiments?" (Haavelmo 1940, p. 59). The answer is no, not by statistics alone.

The reason is that statistics is not sufficient to reveal the complete set of all the potential influences of the measurand. Statistics alone does not lead to exact representations of the measurand. Whatever test-statistics one employs to identify a significant influence, they are all based on the "variance" of that influence. So, to identify a potential influence, variance has to have shown. If not, it will not be detected by any statistical sensor and remain invisible.

Because the available statistics are often not complete enough or sufficient to be used for measurement, recourse has to be taken to other sources of knowledge. The most obvious candidate is theory. However, theories of field phenomena are inexact; they do not provide the complete set of causes that affects the phenomenon in all or some cases, to a considerable or a slight degree.

To complete available statistics and theories, intuition and imagination of experts are needed to run Hempel's "intuitive imaginary experiment." These judgments are by definition not objective and not rational. They are not objective because they are not results of the application of mechanical procedures. They are not rational because they are not mathematically or logically deduced from an already existing explicit model, as in Hempel's "theoretical imaginary experiment." These are "considered judgments" because they include considerations about the most appropriate procedure and model for the measurement at hand.

The involvement of expert judgments, however, creates new problems: Because expert judgments are personal, and (perhaps therefore) experts tend to disagree, the problem is how to arrive at intersubjectivity. Intersubjectivity is acquired by reaching consensus. A related problem is whether each expert is as good as another, and if not how to weigh the expert.

The suggestion proposed in this book is to validate experts in the same way as black-box models, and so actually apply Reichenbach's distinction between the context of discovery and the context of justification: the validation of the expert is not based on an assessment how the expert run his or her thought experiment but on how accurate the outcome of this experiment is. Then the consensus is based on this kind of validation: It is the weighted average of the individual judgments, where weights are determined by the validations of the individual experts.

The problem with the proposed strategy for reaching consensus is that for the validation of experts on specific field phenomena long records (statistics) of experts' past performances are needed. Usually these records do not exist. But instead of taking experts to be individuals, this book proposes to take teams of experts affiliated with a specific scientific institution as the expert agency that should be validated.

The question at the beginning of this book, "What are the rules that make the assignment of numerals to properties of objects or events—that is, measurement—reliable, particular outside the laboratory?" can now be answered in a rather compact way: the main conclusion is that reliable measurement outside the laboratory is clinical measurement, that is, the combination of model-based procedures of attaining precision and calibration combined with a rational consensus of expert judgments. To enable this combination and to validate the measurements as objectively as possible—and thereby increase reliability—the model has to be gray-box designed and validated by structure-oriented behavior tests, and the consensus has to be a weighted aggregation of expert judgments, where the weights of the experts are determined by behavior pattern tests. To make these latter tests appropriate, the expert has to be a scientific institution.

Notes

1. Quoted from a transcript of a US Department of Defense news briefing on 12 February 2002 (Federal News Service 2002). See Beck 2014 for an excellent discussion of these kinds of uncertainties in climate science.

BIBLIOGRAPHY

Ackoff, Russell L. 1974. *Redesigning the Future: A Systems Approach to Societal Problems*. New York: Wiley.

Adair, John G. 1984. The Hawthorne effect: A reconsideration of the methodological artifact. *Journal of Applied Psychology* 69 (2), 334–345.

Adair, John G. 1991. Social cognition, artifact and the passing of the so-called crisis in social psychology. *Canadian Psychology* 32 (3), 445–450.

Aldrich, John. 1989. Autonomy. In *History and Methodology of Econometrics*, ed. N. De Marchi and C. Gilbert, 15–34. Oxford: Clarendon Press.

Aldrich, John. 1994. Haavelmo's identification theory. *Econometric Theory* 10, 198–219.

Andersen, Hans Christian. [1843] 2013. The Nightingale. Fairy Tale no. 64. Translated by Jean Hersholt. Hans Christian Andersen Center. <http://www.andersen.sdu.dk/vaerk/hersholt/TheNightingale_e.html>. (Last accessed on 15-3-2014.)

Anderson, Norman H. 1981. *Foundations of Information Integration Theory*. New York: Academic Press.

Ashby, W. Ross. 1956. *An Introduction to Cybernetics*. London: Chapman and Hall.

Aspinall, W. 2010. A route to more tractable expert advice. *Nature* 463, 21 January, 294–295.

Bapeswara Rao, V.V., and M. Bhaskara Rao. 1992. A three-door game show and some of its variants. *Mathematical Scientist* 17 (2), 89–94.

Barbeau, E. 1991. Fallacies, flaws, and flimflam. *College Mathematics Journal* 22, 307–310.

Bar-Hillel, Maya, and Ruma Falk. 1982. Some teasers concerning conditional probabilities. *Cognition* 11, 109–122.

Barlas, Yaman. 1996. Formal aspects of model validity and validation in system dynamics. *System Dynamics Review* 12 (3), 183–210.

Beck, M. Bruce. 2014. Handling uncertainty in environmental models at the science-policy-society interfaces. In *Error and Uncertainty in Scientific Practice*, ed. M. Boumans, G. Hon, and A. Petersen, 97–135. London: Pickering and Chatto.

Beek, L. 2004. *De Geschiedenis van de Nederlandse Natuurwetenschap* (History of Dutch Natural Science). Kampen: Kok.

Ben-Yehuda, Ori. 2006. Physician judgment in cardiology. The art of medicine lives on. *Journal of the American College of Cardiology* 48 (5), 954–955.

Ben-Yehuda, Ori. 2007. Reply. *Journal of the American College of Cardiology* 49 (9), 1013.

Bjerkholt, O. 2005. Frisch's econometrics laboratory and the rise of Trygve Haavelmo's probability approach. *Econometric Theory* 21, 491–533.

Bogen, James, and James Woodward. 1988. Saving the phenomena. *Philosophical Review* 97 (3), 303–352.

Bohl, Alan H., Matthew J. Liberatore, and Robert L. Nydick. 1995. A tale of two goats . . . and a car, or the importance of assumptions in problem solutions. *Journal of Recreational Mathematics* 27 (1), 1–9.

Boumans, Marcel. 2004. Models in economics. In *The Elgar Companion to Economics and Philosophy*, ed. J. B. Davis, A. Marciano, and J. Runde, 260–282. Northampton, MA: Edward Elgar.

Boumans, Marcel. 2005. *How Economists Model the World into Numbers*. New York: Routledge.

Boumans, Marcel. 2006. The difference between answering a "why"-question and answering a "how much"-question. In *Simulation: Pragmatic Construction of Reality*, ed. J. Lenhard, G. Küppers, and T. Sinn, 107–124. *Sociology of the Sciences Yearbook* 25. Springer.

Boumans, Marcel. 2011. The two-model problem in rational decision making. *Rationality and Society* 23 (3), 371–400.

Boumans, Marcel. 2012. Measurement in economics. In *Handbook of the Philosophy of Science: Philosophy of Economics*, vol. 13, ed. U. Mäki, 395–423. Amsterdam: Elsevier.

Boumans, Marcel, and John B. Davis. 2010. *Economic Methodology: Understanding Economics as a Science*. Basingstoke: Palgrave Macmillan.

Boumans, Marcel, and Giora Hon. 2014. Introduction. In *Error and Uncertainty in Scientific Practice*, ed. M. Boumans, G. Hon, and A. Petersen, 1–12. London: Pickering and Chatto.

Bowman, Raymond T. 1964. Comments on "Qui Numerare Incipit Errare Incipit" by Oskar Morgenstern. *American Statistician* 18 (3), 10–20.

Brockhoff, Klaus. [1975] 2002. The performance of forecasting groups in computer dialogue and face to face discussion. In *The Delphi Method, Techniques and Applications*, ed. H. A. Linstone and M. Turoff, 285–311. Digital version of 1975 edition published by Addison-Wesley. Available at <http://is.njit.edu/pubs/delphibook/>. (Last accessed on 21 August 2014.)

Buddingh, H. 1989. De race met de werkelijkheid: Hoe nauwkeurig zijn de modellen van het Planbureau? (The race with reality: How accurate are the models of the Planbureau?). *NRC-Handelsblad*, 20 April.

Burns, Arthur F., and Wesley C. Mitchell. 1946. *Measuring Business Cycles*. New York: NBER.

Buys Ballot, C. H. D. 1848. Iets over de meteorologische waarnemingen aan het observatorium te Utrecht (On meteorological observations at the observatory at Utrecht). *Algemeene Konst- en Letterbode*, 379–384.

Buys Ballot, C. H. D. 1850. On the great importance of *deviations* from the mean state of the atmosphere for the science of meteorology. *London, Edinburgh and Dublin Philosophical Magazine* 37 (247), 42–49.

Buys Ballot, C. H. D. 1851. *Uitkomsten der Meteorologische Waarnemingen Gedaan in 1849 en 1850 te Utrecht en op Eenige Andere Plaatsen in Nederland* (Results of Meteorological Observations in 1849 and 1850 at Utrecht and at Some Other Locations in The Netherlands). Utrecht: Kemink.

Buys Ballot, C. H. D. 1865. On meteorological observations as made in Holland. *Civil Engineer and Architect's Journal* 28, 245–246.

Buys Ballot, C. H. D. 1872. *Suggestions on a Uniform System of Meteorological Observations*. Utrecht: KNMI.

Buys Ballot, C. H. D. 1882. *Beredeneerd Register op de Werken van het Koninklijk Nederlands Meteorologisch Instituut tot 1882* (Motivated Register of the Works of the Royal Dutch Meteorological Institute till 1882). Utrecht: Kemink.

Camerer, C. 1995. Individual decision making. In *The Handbook of Experimental Economics*, ed. J. H. Kagel and A. Roth, 587–703. Princeton, NJ: Princeton University Press.

Campbell, Norman R. 1920. *Physics: The Elements*. Cambridge: Cambridge University Press.

Campbell, Norman R. 1928. *An Account of the Principles of Measurement and Calculation*. London: Longmans, Green.

Campbell, Norman R. 1940. Notes on physical measurement. *Advancement of Science* 2, 340–342.

Cannon, W. F. 1961. John Herschel and the idea of science. *Journal of the History of Ideas* 22, 215–219.

Cartwright, Nancy. 1983. *How the Laws of Physics Lie*. Oxford: Clarendon Press.

Cartwright, Nancy. 1989. *Nature's Capacities and Their Measurement*. Oxford: Clarendon Press.

Cartwright, Nancy. 1999. *The Dappled World: A Study of the Boundaries of Science*. Cambridge: Cambridge University Press.

Casscells, Ward, Arno Schoenberger, and Thomas Graboys. 1978. Interpretation by physicians of clinical laboratory results. *New England Journal of Medicine* 299 (18), 999–1000.

Chang, Hasok. 2004. *Inventing Temperature: Measurement and Scientific Progress*. Oxford: Oxford University Press.

Christ, Carl F. 1994. The Cowles Commission's Contribution to Econometrics at Chicago, 1939–1955. *Journal of Economic Literature* 32 (1), 30–59.

Churchman, C. W., and P. Ratoosh, eds. 1956. *Measurement. Definitions and Theories*. New York: Wiley.

Cooke, Roger M. 1991. *Experts in Uncertainty: Opinion and Subjective Probability in Science*. New York: Oxford University Press.

Cooke, Roger M. 2009. Resources for the future. Researcher Spotlight Roger Cooke. <http://www.rff.org/Researchers/Pages/ResearcherSpotlightCooke.aspx>. (Last accessed 6 November 2014.)

Cooke, Roger M., and L. H. J. Goossens. 1999. *Procedures Guide for Structured Expert Judgment*. Report EUR 18820. Luxembourg: European Commission.

Cooley, Thomas F., and Edward C. Prescott. 1995. Economic growth and business cycles. In *Frontiers of Business Cycle Research*, ed T. F. Cooley, 1–38. Princeton, NJ: Princeton University Press.

Coombs, Clyde H., Howard Raiffa, and Robert M. Thrall. 1954a. Some views on mathematical models and measurement theory. *Psychological Review* 61 (2), 132–144.

Coombs, Clyde H., Howard Raiffa, and Robert M. Thrall. 1954b. Mathematical models and measurement. In *Decision Processes*, ed. Robert M. Thrall, Clyde H. Coombs, and R. L. Davis, 19–37. New York: Wiley.

Cowles Commission. 1952. *Economic Theory and Measurement: A Twenty Year Research Report, 1932–1952*. Chicago: Cowles Commission for Research in Economics.

Cuhls, K. 2005. Delphi method. In *Delphi Surveys: Teaching material for UNIDO Foresight Seminars*, ed. K. Cuhls, 93–112. Vienna: United Nations Industrial Development Organization.

Dalkey, Norman C. 1969. An experimental study of group opinion: The Delphi method. *Futures* 1 (5), 408–426.

Dalkey, Norman C., and Olaf Helmer. 1963. An experimental application of the Delphi method to the use of experts. *Management Science* 9 (3), 458–467.

Daston, Lorraine. 1995. The moral economy of science. *Osiris* 10, 2–24.

Daston, Lorraine, and Peter Galison. 2007. *Objectivity*. Brooklyn, NY: Zone Books.

de Jong, J., M. Roscam Abbing, and J. Verbruggen. 2010. Voorspellen in Crisistijd: De CPB-ramingen tijdens de Grote Recessie (Forecasting in times of crisis: The CPB forecasts during the Great Recession). CPB Document No. 207. The Hague: CPB.

Dekker, E. 1992. Een procesverbaal van verhoor. *Gewina* 15, 153–162.

den Butter, Frank A.G., and Mary S. Morgan. 1998. What makes the models-policy interaction successful? *Economic Modelling* 15, 443–475.

den Butter Frank A. G., and Mary S. Morgan, eds. 2000. *Empirical Models and Policy-Making: Interaction and Institutions*. London: Routledge.

Don, F. J. H., and J. P. Verbruggen. 2006. Models and methods for economic policy: 60 years of evolution at CPB. *Statistica Neerlandica* 60 (2), 145–170.

Donihue, M. R. 1993. Evaluating the role judgment plays in forecast accuracy. *Journal of Forecasting* 12 (2), 81–92.

Donihue, M. R., and J. Kitchen. 2000. The Troika process, economic models and macroeconomic policy in the USA. In *Empirical Models and Policy-Making: Interaction and Institutions*, ed. F. A. G. den Butter and M. S. Morgan, 229–243. London: Routledge.

Downward, Paul, and Andrew Mearman. 2008. Decision-making at the Bank of England: A critical appraisal. *Oxford Economic Papers* 60, 385–409.

Doyle, Arthur Conan [1902] 2007. *The Hound of the Baskervilles*. London: Penguin.

Eddy, David M. 1982. Probabilistic reasoning in clinical medicine: Problems and opportunities. In *Judgment under Uncertainty: Heuristics and Biases*, ed. Daniel Kahneman, Paul Slovic, and Tversky, 249–267. Cambridge: Cambridge University Press.

Elgin, Catherine Z. 1996. *Considered Judgment*. Princeton, NJ: Princeton University Press.

Eliot, T. S. 1934. *The Rock*. London: Faber & Faber.

Ellis, Brian. 1966. *Basic Concepts of Measurement*. Cambridge: Cambridge University Press.

Erickson, Paul. 2013. "The World the Game Theorists Made." Manuscript.

Evidence-Based Medicine Working Group. 1992. Evidence-based medicine: A new approach to teaching the practice medicine. *Journal of the American Medical Association* 268 (17), 2420–2425.

Federal News Service. 2002. Transcript of the US Department of Defense news briefing on 12 February 2002. <http://www.defense.gov/transcripts/transcript.aspx?transcriptid=2636>. (Last accessed on 21–2–2014.)

Federico, Giovanni, and Antonio Tena. 1991. On the accuracy of foreign trade statistics (1909–1935): Morgenstern revisited. *Explorations in Economic History* 28, 259–273.

Ferguson, Allan, et al. 1940. Quantitative estimates of sensory events. *Advancement of Science* 2, 331–349.

Finkelstein, Ludwik. 1975. Fundamental concepts of measurement: Definition and scales. *Measurement and Control* 8 (3): 105–111.

Finkelstein, Ludwik. 1982. Theory and philosophy of measurement. In *Handbook of Measurement Science. Volume 1 Theoretical Fundamentals*, ed. P. H. Sydenham, 1–29. Chichester: Wiley.

Fleischacker, Samuel. 1999. *A Third Concept of Liberty: Judgment and Freedom in Kant and Adam Smith*. Princeton, NJ: Princeton University Press.

Franses, Philip Hans. 2008. Merging models and experts. *International Journal of Forecasting* 24, 31–33.

Franses, Philip Hans, Henk C. Kranendonk, and Debby Lanser. 2007. On the optimality of expert-adjusted forecasts. CPB Discussion Paper No. 92. The Hague: CPB.

Franses, Philip H., Henk C. Kranendonk, and Debby Lanser. 2011. One model and various experts: Evaluating Dutch macroeconomic forecasts. *International Journal of Forecasting* 27, 482–495.

Friedman, Daniel. 1998. Monty Hall's three doors: Construction and deconstruction of a choice anomaly. *American Economic Review* 88 (4), 933–946.

Frigerio, Aldo, Alessandro Giordani, and Luca Mari. 2010. Outline of a general model of measurement. *Synthese* 175, 123–149.

Frisch, Ragnar. 1926. Sur un problème d'économie pure. *Norsk Matematisk Forenings Skrifter* 1 (16), 1–40. Translated by J. S. Chipman as "On a problem in pure economics." In *Preferences, Utility, and Demand*, ed. J. S. Chipman, L. Hurwicz, M. K. Richter and H. F. Sonnenschein. New York: Harcourt Brace Jovanovich, 1971.

Frisch, Ragnar. 1934. *Statistical Confluence Analysis by Means of Complete Regression Systems*. Oslo: Universitetets Økonomiske Institutt.

Frisch, Ragnar. 1938. The cyclical intensity coefficient of one variate with respect to another. Unpublished letter, dated 15 September 1938. Archives of the League of Nations, Geneva.

Galilei, Galileo. [1632] 1967. *Dialogue Concerning the Two Chief World Systems: Ptolemaic and Copernican*. Translated by Stillman Drake. 2nd ed. Berkeley: University of California Press.

Galton, Francis. 1879. Psychometric experiments. *Brain* 2 (2),149–162.

Gauss, Carl Friedrich. 1809. *Theoria Motus Corporum Coelestium in Sectionibus Conicis Solem Ambientium*. Hamburg: Friedrich Perthes and I.H. Besser.

Gauss, Carl Friedrich. [1821] 1995. *Theory of the Combination of Observations: Least Subject to Error. Part One, Part Two, Supplement*. Trans. G. W. Stewart. Philadelphia: Society for Industrial and Applied Mathematics.

Gillman, Leonard. 1992. The car and the goats. *American Mathematical Monthly* 99 (1), 3–7.

Goossens, L. H. J., R. M. Cooke, A. H. Hale, and Lj. Rodic-Wiersma. 2008. Fifteen years of expert judgment at TU Delft. *Safety Science* 46, 234–244.

Granberg, Donald, and Thad A. Brown. 1995. The Monty Hall dilemma. *Personality and Social Psychology Bulletin* 21 (7), 711–723.

Guyatt, G. H., A. D. Oxman, G. E. Vist, R. Kunz, Y. Flack-Ytter, P. Alonso-Coello, and J. H. J. Schünemann. 2008. Rating quality of evidence and strength of recommendations: GRADE: An emerging consensus on rating quality of evidence and strength of recommendations. *British Medical Journal* 336 (7650), 924–926.

Haavelmo, Trygve. 1938. Drei Beispiele der ökonometrischen Forschung in den Niederlanden. *Weltwirtschaftliches Archiv* 48, 7*–11*.

Haavelmo, Trygve. 1940. The problem of testing economic theories by means of passive observations. In *Report of Sixth Annual Research Conference on Economics and Statistics at Colorado Springs, July 1–26, 1940*, 58–60. Cowles Commission for Research in Economics.

Haavelmo, Trygve. 1941a. "On the Theory and Measurement of Economic Relations." Cambridge, MA. Available at <http://www.sv.uio.no/econ/english/about/facts-and-figures/nobel-prize-winners/haavelmo/published-works/. TH1941c>. (Last accessed on 14–3–2014.)

Haavelmo, Trygve. 1941b. The effect of the rate of interest on investment: A note. *Review of Economic Statistics* 23 (1), 49–52.

Haavelmo, Trygve. 1944. *The Probability Approach in Econometrics*. Supplement to *Econometrica* 12.

Haavelmo, Trygve. [1939] 2008. Om statistisk "testing" av hypoteser i den økonomiske teori, the Third Scandinavian Meeting for Younger Economists in Copenhagen May 27–30 1939, Aarhus. Translated by Erik Biørn as "On the statistical 'testing' of hypotheses in economic theory." Available at <http://www.sv.uio.no/econ/english/about/facts-and-figures/nobel-prize-winners/haavelmo/published-works/. TH1939e>. (Last accessed on 14–3–2014.)

Hald, A. 1986. Galileo's statistical analysis of astronomical observations. *International Statistical Review* 54 (2), 211–220.

Harper, Douglas. 2001–2014. Online Etymology Dictionary. <http://www.etymonline.com/>. (Last accessed on 15–3–2014.)

Harrison, Glenn W., and John A. List. 2004. Field experiments. *Journal of Economic Literature* 42 (4), 1009–1055.

Hausman, Daniel M. 1992. *The Inexact and Separate Science of Economics*. Cambridge: Cambridge University Press.

Heidelberger, Michael. 1994. Three strands in the history of the representational theory of measurement. Working paper, Humboldt University, Berlin.

Helmer, Olaf. 1983. *Looking Forward: A Guide to Futures Research*. Beverly Hills, CA: Sage.

Helmer, Olaf, and Nicholas Rescher. 1958. On the epistemology of the inexact sciences. P-1513, October 13. Rand.

Helmer, Olaf, and Nicholas Rescher. 1959. On the epistemology of the inexact sciences. *Management Science* 6 (1), 25–52.

Helmholtz, Hermann von. 1887. Zählen und Messen, erkenntnis-theoretisch betrachtet. *Philosophische Aufsätze*. Leipzig: Fues.

Hempel, Carl G. 1945. Geometry and empirical science. *American Mathematical Monthly* 52 (1), 7–17.

Hempel, Carl G. 1952. Symposium: Problems of concept and theory formation in the social sciences. In *Science, Language, and Human Rights*, 65–86. Philadelphia: University of Pennsylvania Press.

Henderson, Robert. 1924. A new method of graduation. *Transactions* 25, 29–39.

Henderson, Robert. 1925. Further remarks on graduation. *Transactions* 26, 52–57.

Henderson, Robert. [1918] 1938. *Mathematical Theory of Graduation*. 2nd ed. New York: Actuarial Society of America.

Hendry, David F., and Mary S. Morgan, eds. 1995. *The Foundations of Econometric Analysis*. Cambridge: Cambridge University Press.

Herschel, John F. W. 1830. *Preliminary Discourse on the Study of Natural Philosophy*. New ed. London: Longman.

Herschel, John F. W. 1836. Instructions for making and registering meteorological observations at various stations in Southern Africa and other countries in the south seas, as also at sea. *Edinburgh New Philosophical Journal* 21, 135–149.

Heukelom, F. 2010. Measurement and decision making at the University of Michigan in the 1950s and 1960s. *Journal of the History of the Behavioral Sciences* 46 (2), 189–207.

Hölder, O. L. 1901. Die Axiome der Quantität und die Lehre vom Mass. *Berichte über die Verhandlungen der Königlich Sächsischen Gesellfschaft der Wissenschaften zu Leipzig, Mathematisch-Physikaliche Classe* 53, 1–64.

Hon, Giora. 1989. Towards a typology of experimental errors: An epistemological view. *Studies in History and Philosophy of Science* 20 (4), 469–504.

Hoover, Kevin D. 1994. Econometrics as observation: The Lucas critique and the nature of econometric inference. *Journal of Economic Methodology* 1 (1), 65–80.

Hoover, Kevin D. 2001. *Causality in Macroeconomics*. Cambridge: Cambridge University Press.

Hoover, Kevin D. 2002a. Symposium on Marshall's tendencies: 5 Sutton's critique of econometrics. *Economics and Philosophy* 18 (1), pp. 45–54.

Hoover, Kevin D. 2002b. Econometrics and reality. In *Fact and Fiction in Economics*, ed. U. Mäki, 152–177. Cambridge: Cambridge University Press.

Information Services Department of the Library of the Health Sciences-Chicago, University of Illinois at Chicago. 2006. *Levels of Evidence: Evidence Based Practice in the Health Sciences: Evidence-Based Nursing Tutorial.*

Innocenti, Alessandro, and Carlo Zappia. 2005. Thought- and performed experiments in Hayek and Morgenstern. In *The Experiment in the History of Economics*, ed. Philippe Fontaine and Robert Leonard, 71–97. Oxon: Routledge.

Joint Committee for Guides in Metrology (JCGM) 100. 2008. *Evaluation of Measurement Data: Guide to the Expression of Uncertainty in Measurement*. JCGM.

JCGM 104. 2009. *Evaluation of Measurement Data: An Introduction to the "Guide to the Expression of Uncertainty in Measurement" and Related Documents*. JCGM.

JCGM 200. 2012. *International Vocabulary of Metrology: Basic and General Concepts and Associated Terms*, 3rd edition. JCGM.

Jick, Todd D. 1979. Mixing qualitative and quantitative methods: Triangulation in action. *Administrative Science Quarterly* 24, 602–611.

Kahl, Russell, ed. 1971. *Selected Writings of Hermann von Helmholtz*. Middletown, CT: Wesleyan University Press.

Kahneman, Daniel, Paul Slovic, and Amos Tversky, eds. 1982. *Judgment under Uncertainty: Heuristics and Biases*. Cambridge: Cambridge University Press.

Kant, Immanuel. [1892] 1914. *The Critique of Judgement*. Trans. J. H. Bernard. 2nd ed. London: Macmillan.

Karthikeyan, Ganesan. 2007. Evidence-based medicine and clinical judgment: An imaginary divide. *Journal of the American College of Cardiology* 49 (9), 1012.

Karthikeyan, Ganesan, and P. Pais. 2010. Clinical judgement and evidence-based medicine: Time for reconciliation. *Indian Journal of Medical Research* 132 (5), 623–626.

Kendall, Maurice G., and Alan Stuart. 1963. *The Advanced Theory of Statistics*. 2nd ed. Vol. 1. London: Charles Griffin.

Keynes, John Maynard. 1939. Professor Tinbergen's method. *Economic Journal* 49 (195), 558–568.

Keynes, J. M. 1940. Comment. *Economic Journal* 50, 154–156.

Kim, J., N. De Marchi, and M. S. Morgan. 1995. Empirical model particularities and belief in the natural rate hypothesis. *Journal of Econometrics* 67, 81–102.

Klein, Judy L. 1997. *Statistical Visions in Time: A History of Time Series Analysis, 1662–1938*. Cambridge: Cambridge University Press.

Kluger, B., and D. Friedman. 2010. Financial engineering and rationality: Experimental evidence based on the Monty Hall problem. *Journal of Behavioral Finance* 11 (1), 31–49.

KNMI. 1954. *Koninklijk Nederlands Meteorologisch Instituut 1854–1954* (Royal Meteorological Institute, 1854–1954). Den Haag: Staatsdrukkerij- en Uitgeverijbedrijf.

Koopmans, Tjalling C. 1937. *Linear Regression Analysis of Economic Time Series*. Netherlands Economic Institute, Publication No. 20. Haarlem: Bohn.

Koopmans, Tjalling C. 1941. The logic of econometric business-cycle research. *Journal of Political Economy* 49 (2), 157–181.

Koopmans, Tjalling C. 1947. Measurement without theory. *Review of Economics and Statistics* 29 (3), 161–172.

Koopmans, Tjalling C. 1951. Comments on causality and identification. Cowles Commission discussion paper: Statistics no. 359, 21 March 1951. Available at <http://cowles.econ.yale.edu/P/ccdp/st/s-0359.pdf>. (Last accessed on 14–3–2014.)

Koopmans, Tjalling C., Herman Rubin, and Roy B. Leipnik. 1950. Measuring the equation systems of dynamic economics. In *Statistical Inference in Dynamic Economic Models*, Cowles Commission Monograph 10, ed. T. C. Koopmans, 53–237. New York: Wiley.

Krantz, David H., R. Duncan Luce, Patrick Suppes, and Amos Tversky. 1971. *Foundations of Measurement*. Vol. 1: *Additive and Polynomial Representations*. New York: Academic Press.

Krantz, David H., R. Duncan Luce, Patrick Suppes, and Amos Tversky. 1989. *Foundations of Measurement*. Vol. 2: *Geometrical, Threshold and Probabilistic Representations*. New York: Academic Press.

Krantz, David H., R. D. Luce, P. Suppes, and A. Tversky. 1990. *Foundations of Measurement*. Vol. 3: *Representation, Axiomatization, and Invariance*. New York: Academic Press.

Kuhn, Harold W. 2004. Introduction. In *Theory of Games and Economic Behavior*, by John von Neumann and Oskar Morgenstern, vii–xiv. 60th Anniversary Edition. Princeton, NJ: Princeton University Press.

Kuhn, Thomas S. 1970. *The Structure of Scientific Revolutions*. 2nd ed. International Encyclopedia of Unified Science, vol. 2, no. 2. Chicago: University of Chicago Press.

Kuklick, Henrika, and Robert E. Kohler. 1996. Introduction to "Science in the Field." *Osiris* 11, 1–14.

Kuznets, Simon. 1950a. Conditions of statistical research. *Journal of the American Statistical Association* 45 (249), 1–14.

Kuznets, Simon. 1950b. Review of *On the Accuracy of Economic Observations*. *Journal of the American Statistical Association* 45, 576–579.

Laesecke, A. 2002. Through measurement to knowledge: The inaugural lecture of Heike Kamerlingh Onnes (1882). *Journal of Research of the National Institute of Standards and Technology* 107 (3), 261–277.

Lambert, R. 2005. Inside the MPC. *Bank of England Quarterly Bulletin* 45 (1), 56–65.

Landré, Corneille L. 1881. Over de functie φ van de methode der kleinste kwadraten (On the function φ of the method of least squares). *Nieuw Archief voor Wiskunde* 7, 214–219.

Landré, Corneille L. 1884. De middelbare fout bij waarneming ter bepaling van meer dan een onbekende (The mean error of observation to determine more than one unknown). *Nieuw Archief voor Wiskunde* 10, 1–17.

Landré, Corneille L. 1889. Over correctie van getalreeksen door middel van tweede verschillen (On the correction of time series by second differences). *Nieuw Archief voor Wiskunde* 15, 15–22.

Landré, Corneille L. 1900a. Afronding van sterftecijfers zonder een wet van afsterving ten grondslag te leggen (Graduation of mortality rates without founding it on a law of mortality). *Archief voor de Verzekerings-Wetenschap en Aanverwante Vakken* 4, 314–358.

Landré, Corneille L. 1900b. Afronding naar een nieuw beginsel (Graduation to a new principle). *Archief voor de Verzekerings-Wetenschap en Aanverwante Vakken* 4, 471–502.

Landré, Henriette F. 1906. Corneille L. Landré door zijne dochter. *Jaarboekje voor 1906 uitgegeven door de Vereeniging voor Levensverzekering*, 194–208.

Laplace, Pierre-Simon. 1809. Mémoire sur les approximations des formules qui sont functions de très grands nombres et sur leur application aux probabilities. *Mémoires de l'Académie des Sciences de Paris*, 353–415, 559–565.

Larkey, Patrick D., Richard A. Smith, and Joseph B. Kadane. 1989. It's okay to believe in the "hot hand." *Chance* 2, 22–30.

Lawrence, Michael, Paul Goodwin, Marcus O'Connor, and Dilek Önkal. 2006. Judgmental forecasting: A review of progress over the last 25 years. *International Journal of Forecasting* 22, 493–518.

Legendre, Adrien-Marie. 1805. *Novelles methods pour la determination des orbites des comètes*. Paris.

Lehrer, Keith, and Carl Wagner. 1981. *Rational Consensus in Science and Society: A Philosophical and Mathematical Study*. Dordrecht: Reidel.

Leonard, Robert. 2010. *Von Neumann, Morgenstern, and the Creation of Game Theory: From Chess to Social Science, 1900–1960*. Cambridge: Cambridge University Press.

London, Dick. 1985. *Graduation: The Revision of Estimates*. Winsted: ACTEX.

Maistrov, L. E. 1974. *Probability Theory: A Historical Sketch*. Trans. and ed. S. Kotz. New York: Academic.

Mach, Ernst. 1896. *Die Principien der Wärmelehre*. Leipzig: Barth.

Mansnerus, Erika. 2014. *Modelling for Policy. How Mathematical Techniques Keep Us Healthy*. New York: Palgrave Pivot.

Manuel, Frank E. 1974. *The Religion of Isaac Newton: The Fremantle Lectures 1973*. Oxford: Clarendon Press.

Marget, Arthur W. 1929. Morgenstern on the methodology of economic forecasting. *Journal of Political Economy* 37 (3), 312 339.

Marshall, Alfred. [1890] 1920. *Principles of Economics*. 8th ed. London: Macmillan.

Martini, Carlo. 2014. Experts in science: A view from the trenches. *Synthese* 191 (1), 3–15.

Martini Carlo, and M. Boumans, eds. 2014. *Experts and Consensus in Social Science*. New York: Springer.

Maxwell, James Clerk. 1873. *A Treatise of Electricity and Magnetism*. Oxford: Clarendon Press.

Maxwell, James Clerk [1856] 1965. On Faraday's lines of force. In *The Scientific Papers of James Clerk Maxwell*, vol. 1, ed. W. D. Niven, 155–229. New York: Dover.

Michell, Joel. 1993. The origins of the representational theory of measurement: Helmholtz, Hölder, and Russell. *Studies in History and Philosophy of Science* 24 (2), 185–206.

Michell, Joel. 1999. *Measurement in Psychology: A Critical History of a Methodological Concept*. Cambridge: Cambridge University Press.

Michell, Joel, and C. Ernst. 1996. The axioms of quantity and the theory of measurement, part I. *Journal of Mathematical Psychology* 40, 235–252.

Michell, Joel, and C. Ernst. 1997. The axioms of quantity and the theory of measurement, part II. *Journal of Mathematical Psychology* 41, 345–356.

Mill, John Stuart. [1843] 1911. *A System of Logic, Ratiocinative and Inductive*. 8th ed. London: Longmans, Green.

Miller, Morton D. 1946. *Elements of Graduation*. Philadelphia: Actuarial Society of America and American Institute of Actuaries.

Mirowski, Philip. 1992. What were von Neumann and Morgenstern trying to accomplish? In *Towards a History of Game Theory*, ed. E. Roy Weintraub. *History of Political Economy* 24 (supplement), 113–147.

Mood, A. M., F. A. Graybill, and D. C. Boes. 1974. *Introduction to the Theory of Statistics*. 3rd ed. Tokyo: McGraw-Hill.

Morgan, Mary S. 1990. *The History of Econometrics Ideas*. Cambridge: Cambridge University Press.

Morgan, Mary S. 2000. The relevance of economic modelling for policy-making: A panel discussion. In *Empirical Models and Policy-Making: Interaction and Institutions*, ed. F. A. G. den Butter and M. S. Morgan, 259–278. New York: Routledge.

Morgan, Mary S. 2005. Experiments versus models: New phenomena, inference and surprise. *Journal of Economic Methodology* 12 (2), 317–329.

Morgan, Mary S. 2012. *The World in the Model: How Economists Work and Think*. Cambridge: Cambridge University Press.

Morgan, J. P., N. R. Chaganty, R. C. Dahiya, and M. J. Doviak. 1991a. Let's make a deal: The player's dilemma. *American Statistician* 45 (4), 284–287.

Morgan, J. P., N. R. Chaganty, R. C. Dahiya, and M. J. Doviak. 1991b. Rejoinder to vos Savant. *American Statistician* 45 (4), 347–348.

Morgenstern, Oskar. 1928. *Wirtschaftsprognose: Eine Untersuchung ihrer Voraussetzungen und Möglichkeiten*. Vienna: Springer.

Morgenstern, Oskar. 1950. *On the Accuracy of Economic Observations*. Princeton, NJ: Princeton University Press.

Morgenstern, Oskar. 1951. Prolegomena to a Theory of Organisation. RM-734, December 10, Rand.

Morgenstern, Oskar. 1954. Experiment and large scale computation in economics. In *Economic Activity Analysis*, ed. Oskar Morgenstern, 484–549. New York: Wiley.

Morgenstern, Oskar. 1959. *International Financial Transactions and Business Cycles*. National Bureau of Economic Research Book Series in Business Cycles. Princeton, NJ: Princeton University Press.

Morgenstern, Oskar. 1963a. *On the Accuracy of Economic Observations*. 2nd ed. Princeton, NJ: Princeton University Press.

Morgenstern, Oskar. 1963b. Qui numerare incipit errare incipit. *Fortune* 68, 142–144, 173–174, 178, 180.

Morgenstern, Oskar. 1966. Nature's Attitude and Rational Behavior. Econometric Research Program Research Paper 13, Princeton University.

Morrison, Margaret, and Mary S. Morgan. 1999. Models as mediating instruments. In *Models as Mediators*, ed. M. S. Morgan and M. Morrison, 10–37. Cambridge: Cambridge University Press.

Moscati, I. 2013. Were Jevons, Menger, and Walras really cardinalists? On the notion of measurement in utility theory, psychology, mathematics, and other disciplines, 1870–1910. *History of Political Economy* 45 (3), 373–414.

Mounier, Guillaume J. D. 1903. Iets over de grondslagen van de methode der kleinste kwadraten (On the foundations of the method of least squares). *Archief voor de Verzekerings-Wetenschap and Aanverwante Vakken* 6, 1–43.

Mounier, Guillaume J. D. 1906a. In memoriam. Corneille Louis Landré. *Archief voor de Verzekerings-Wetenschap en Aanverwante Vakken* 8, 227–247.

Mounier, Guillaume J. D. 1906b. Veelvuldig voorkomende toepassingen van de method der kleinste kwadraten (Frequently occurring applications of the method of least squares). *Archief voor de Verzekerings-Wetenschap en Aanverwante Vakken* 8, 309–348.

Nalebuff, Barry. 1987. Puzzles: Choose a curtain, duel-ity, two point conversions, and more. *Journal of Economic Perspectives* 1, 157–163.

Nelson, Y., and S. Stoner. 1996. Results of the Delphi VIII Survey of Oil Prices Forecasts. California Energy Commission Staff Report, P300-395-017B.

Nesbitt, C. J. 1989. Personal reflections on actuarial science in North America from 1900. In *A Century of Mathematics in America, Part III*, ed. P. Duren, 617–638. Providence, RI: American Mathematical Society.

Nichols, Charles K. 1942. The statistical work of the League of Nations in economics, financial, and related fields. *Journal of the American Statistical Association* 37 (219), 336–342.

Orne, Martin T. 1973. Communication by the total experimental situation: Why it is important, how it is evaluated, and its significance for the ecological validity of findings. In *Communication and Affect*, ed. P. Pliner, L. Krames, and T. Alloway, 157–191. New York: Academic Press.

Pais, Abraham. 1982. *"Subtle Is the Lord . . .": The Science and the Life of Albert Einstein*. Oxford: Oxford University Press.

Paraira, M. C. 1905–1907. Corneille Louis Landré (1838–1905). *Nieuw Archief voor Wiskunde* 7, 1–6.

Pauker, Stephen G., and Jerome P. Kassirer. 1980. The threshold approach to clinical decision making. *New England Journal of Medicine* 302 (20), 1109–1117.

Pereira, Alexandre C., and Whady Hueb. 2007. Reply. *Journal of the American College of Cardiology* 49 (9), 1012–1013.

Pereira, Alexandre C., N. H. M. Lopes, P. R. Soares, J. E. Krieger, S. A. de Oliveira, L. A. M. Cesar, J. A. F. Ramires, and W. Hueb. 2006. Clinical judgment and treatment options in stable multivessel coronary artery disease. *Journal of the American College of Cardiology* 48 (5), 948–953.

Persons, Warren M. 1924. Some fundamental concepts of statistics. *Journal of the American Statistical Association* 19 (145), 1–8.

Pfanzagl, J. 1968. *Theory of Measurement*. Würzburg: Physica Verlag.

Poincaré, Henri. [1914] 1996. *Science and Method.* Trans. F. Maitland. Bristol, UK: Thoemmes Press.

Polanyi, Michael. 1946. *Science, Faith and Society*. London: Oxford University Press.

Popper, Karl. 1979. Three worlds. *Michigan Quarterly Review* 18 (1), 1–23.

Porter, Theodore M. 1995. *Trust in Numbers: The Pursuit of Objectivity in Science and Objective Life.* Princeton, NJ: Princeton University Press.

Qin, Duo. 1993. *The Formation of Econometrics: A Historical Perspective*. Oxford: Clarendon Press.

Raiffa, Howard, and R. Duncan Luce. 1957. *Games and Decisions: Introduction and Critical Survey.* New York: Wiley, London: Chapman.

Reiss, Julian. 2014. Struggling over the soul of economics: Objectivity vs expertise. In *Experts and Consensus in Social Science*, ed. C. Martini and M. Boumans, 131–152. New York: Springer.

Rescher, Nicholas. 2006. The Berlin school of logical empiricism and its legacy. *Erkenntnis* 64, 281–304.

Roberts, F. S. 1979. *Measurement Theory with Applications to Decisionmaking, Utility, and the Social Sciences*. London: Addison-Wesley

Rodgers, P. 1997. Changes at the Bank of England. *Bank of England Quarterly Bulletin* 37 (3), 241–247.

Rosenblueth, Arturo, and Norbert Wiener. 1945. The role of models in science. *Philosophy of Science* 12 (4), 316–321.

Roth, Alvin E. 1993. The Early History of Experimental Economics. *Journal of the History of Economic Thought* 15, 184–209.

Sackett, D. L. 2000. The sins of expertness and a proposal for redemption. *British Journal of Medicine* 320 (7244), 1283.

Sackett, D. L., S. E. Straus, W. S. Richardson, W. Rosenberg, and R. B. Haynes. 2000. *Evidence-Based Medicine: How to Practice and Teach EBM*. 2nd ed. Edinburgh: Churchill Livingstone.

Samuelson, Paul A. 1967. *Economics: An Introductory Analysis*. 7th ed. New York: McGraw-Hill.

Savage, C. W., and P. Ehrlich. 1992. A brief introduction to measurement theory and to the essays. In *Philosophical and Foundational Issues in Measurement Theory*, ed. C. W. Savage and P. Ehrlich, 1–14. Hillsdale, NJ: Lawrence Erlbaum Associates.

Savage, Leonard J. 1954. *The Foundations of Statistics*. New York: John Wiley and Sons.

Scherokman, Barbara. 1997. Selecting and interpreting diagnostic tests. *Permanente Journal* 1 (2), 4–7.

Schwarz, Astrid, and Wolfgang Krohn. 2011. Experimenting with the concept of experiment: Probing the epochal break. In *Science Transformed? Debating Claims of an Epochal Break*, ed. A. Nordmann, H. Radder. and G. Schiemann, 119–134. Pittsburgh: University of Pittsburgh Press.

Selvin, Stevin. 1975. A problem in probability. *American Statistician* 29, 67.

Seymann, Richard G. 1991. Comment. *American Statistician* 45 (4), 287–288.

Seymann, Richard G. 1992. Response. *American Statistician* 46 (3), 242–243.

Simon, Herbert A. 1950. The causal principle and the identification problem. Cowles Commission discussion paper: Statistics no. 353, 19 December 1950. Available at <http://cowles.econ.yale.edu/P/ccdp/st/s-0353.pdf>. (Last accessed on 14–3–2014.)

Simon, Herbert A. 1953. Causal ordering and identifiability. In *Studies in Econometric Method*, ed. W. C. Hood and T. C. Koopmans, 49–74. Cowles Commission for Research in Economics, monograph no. 14. New York: Wiley.

Simpson, Thomas. 1755. A letter to the right honourable George Earl of Macclesfield, president of the Royal Society, on the advantage of taking the mean of a number of observations, in practical astronomy. *Philosophical Transactions* 49, 82–93.

Solow, Robert M. 1970. *Growth Theory: An Exposition*. Oxford: Clarendon Press.

Stevens, Stanley Smith. 1939. Psychology and the science of science. *Psychological Bulletin* 36 (4), 221–263.

Stevens, Stanley Smith. 1946. On the theory of scales of measurement. *Science* 103 (2684), 677–680.

Stevens, Stanley Smith. 1956. Measurement, psychophysics and utility. In *Measurement. Definitions and Theories*, ed. C. W. Churchman and P. Ratoosh, 18–63. New York: Wiley.

Stevens, Stanley Smith, and Hallowell Davis. 1938. *Hearing: Its Psychology and Physiology*. New York: Wiley.

Stewart, G. W. 1995. Translator's introduction. In *Theory of the Combination of Observations: Least Subject to Error. Part One, Part Two, Supplement*, by Carl Friedrich Gauss, trans. G. W. Stewart, ix–xi. Philadelphia: Society for Industrial and Applied Mathematics.

Stigler, Stephen M. 1978. Mathematical statistics in the early states. *Annals of Statistics* 6 (2), 239–265.

Stigler, Stephen M. 1986. *The History of Statistics: The Measurement of Uncertainty before 1900*. Cambridge, MA: Harvard University Press.

Suppes, Patrick, and J. L. Zinnes. 1963. Basic measurement theory. In *Handbook of Mathematical Psychology*, ed. R. Duncan Luce, Robert R. Bush, and Eugene Galanter, vol. 1, 1–76. New York: Wiley.

Sutton, John. 2000. *Marshall's Tendencies. What Can Economists Know?* Leuven: Leuven University Press, Cambridge: MIT Press.

Swijting, Z. G. 1987. The objectification of observation: Measurement and statistical methods in the nineteenth century. In *The Probabilistic Revolution*, vol. 1: *Ideas in History*, ed. L. Krüger, L. J. Daston, and M. Heidelberger, 261–285. Cambridge, MA: MIT Press.

Sydenham, Peter H. 1979. *Measuring Instruments: Tools of Knowledge and Control*. Stevenage: Peter Peregrinus.

Tarski, A. 1954. Contributions to the theory of models, I, II. *Indagationes Mathematicae* 16, 572–588.

Thomson, W. 1889. Electrical units of measurement. In *Popular Lectures and Adresses*, vol. 1: *Constitution of Matter*, 73–136. London: Macmillan.

Thrall, Robert M., Clyde H. Coombs, and Robert L. Davis, eds. 1954. *Decision Processes*. New York: Wiley.

Tinbergen, Jan. 1936a. Kan hier te lande, al dan niet na overheidsingrijpen, een verbetering van de binnenlandse conjunctuur intreden, ook zonder verbetering van onze exportpositie? (In this country, whether or not after government intervention, could there be an improvement in the domestic economy, even without improving our export position?) *Prae-adviezen voor de Vereeniging voor de Staathuishoudkunde en de Statistiek* (Pre-advices for the Society of Economics and Statistics). The Hague: Nijhoff.

Tinbergen, Jan. 1936b. Memorandum on the continuation of the League's business cycle research in a statistical direction. unpublished memorandum. Archive of the League of Nations, Geneva.

Tinbergen, Jan. 1939a. *A Method and Its Application to Investment Activity. Statistical Testing of Business-Cycle Theories*, vol. 1. Geneva: League of Nations.

Tinbergen, Jan. 1939b. *Business Cycles in the United States of America, 1919–1932. Statistical Testing of Business-Cycle Theories*, vol. 2. Geneva: League of Nations.

Tinbergen, Jan. 1940. On a method of statistical business-cycle research: A reply. *Economic Journal* 50, 141–154.

Tinbergen, Jan. 1979. Recollections of professional experiences. *Banca Nazionale del Lavoro Quarterly Review* 32 (131), 331–360.

Tinbergen, Jan. 1987. Over modellen (On models). In *Lessen uit het Verleden: 125 jaar Vereniging voor de Staathuishoudkunde* (Lessons from the Past: 125-Year Society of Economics), ed. A. Knoester, 99–112. Leiden: Stenfert Kroese.

Tinbergen, Jan. [1982] (2003). The need of a synthesis. In *Jan Tinbergen: The Centennial Volume,* ed. J. Kol, 303–306. Rotterdam University. Translation of "De noodzaak van een synthese." *Economisch Statistische Berichten,* 1 December 1982, 1284–1285.

Tversky, Amos, and Thomas Gilovich. 1989. The cold facts about the "hot hand" in basketball. *Chance* 2, 16–21.

Tversky, Amos, and Daniel Kahneman. 1974. Judgment under uncertainty: Heuristics and biases. *Science,* New Series 185 (4157), 1124–1131.

van Everdingen, Ewoud. 1953. *C.H.D. Buys Ballot 1817–1890.* The Hague: Daamen.

van Lunteren, F. 1998. De oprichting van het Koninklijk Nederlands Meteorologisch Instituut: Humboldtiaanse wetenschap, international samenwerking en praktisch nut (The foundation of the Royal Dutch Meteorological Institute: Humboldtian science, international cooperation, and practical use). *Gewina* 21, 216–243.

Verbruggen, J. 2011. De 4 "ellen" van het voorspellen: SAFFIER II. CPB presentation. <http://www.cpb.nl/publicatie/presentatie-de-4-ellen-van-het-voorspellen-saffier-ii>. (Last accessed on 17–3–2014.)

Vining, Rutledge. 1949. Koopmans on the choice of variables to be studied and of methods of measurement. *Review of Economics and Statistics* 31 (2), 77–86.

von Neumann, John, and Oskar Morgenstern. 1944. *Theory of Games and Economic Behavior.* Princeton, NJ: Princeton University Press.

von Neumann, John, and Oskar Morgenstern 1953. *Theory of Games and Economic Behavior.* 3rd ed. Princeton, NJ: Princeton University Press.

von Neumann, John, and Oskar Morgenstern. 2004. *Theory of Games and Economic Behavior. Sixtieth Anniversary Edition.* Princeton, NJ: Princeton University Press.

vos Savant, Marilyn. 1990. Ask Marilyn. *Parade Magazine,* September 9, 13.

vos Savant, Marilyn. 1991. Marilyn vos Savant's reply. *American Statistician* 45 (4), 347.

Warwick, Andrew. 1995. The laboratory of theory or what's exact about the exact sciences? In *The Values of Precision,* ed. M. N. Wise, 311–351. Princeton, NJ: Princeton University Press.

Wendell, John P. 1992. Comment. *American Statistician* 46 (3), 242.

Whewell, William. [1847] 1967. *The Philosophy of Inductive Sciences, Part Two.* Reprint of the 2nd ed. New York: Johnson Reprint Corp.

White, K. Preston. 1999. Systems design. In *Handbook of Systems Engineering and Management,* ed. A. P. Sage and W. B. Rouse, 455–481. New York: Wiley.

Whittaker, Edmund T. 1922. On a new method of graduation. *Proceedings of the Edinburgh Mathematical Society* 41, 63–75.

Whittaker, Edmund T., and George Robinson. 1924. *The Calculus of Observations: A Treatise on Numerical Mathematics.* London: Blackie.

Whittaker, Edmund T., and George Robinson 1944. *The Calculus of Observations: A Treatise on Numerical Mathematics.* 4th ed. London: Blackie.

Wilson, Edwin B. 1946. The probability approach to econometrics by Trygve Haavelmo. *Review of Economics and Statistics* 28 (3), 173–174.

INDEX